Advanced Information and Knowledge Processing

Series Editors

Professor Lakhmi Jain
Lakhmi.jain@unisa.edu.au
Professor Xindong Wu
xwu@cs.uvm.edu

For other titles published in this series, go to
www.springer.com/series/4738

Dorothy Monekosso · Paolo Remagnino
Yoshinori Kuno

Editors

Intelligent Environments

Methods, Algorithms and Applications

 Springer

Editors
Dorothy Monekosso, PhD
Kingston University, UK

Paolo Remagnino, PhD
Kingston University, UK

Yoshinori Kuno, PhD
Saitama University, Japan

AI&KP ISSN: 1610-3947
ISBN: 978-1-84800-345-3   e-ISBN: 978-1-84800-346-0
DOI: 10.1007/978-1-84800-346-0

British Library Cataloguing in Publication Data
A catalogue record for this book is available from the British Library

Library of Congress Control Number: 2008934654

Printed on acid-free paper

Springer Science+Business Media
springer.com

To our parents.

Preface

Relatively new research fields such as ambient intelligence, intelligent environments, ubiquitous computing, and wearable devices have emerged in recent years. These fields are related by a common theme: making use of novel technologies to enhance user experience by providing user-centric intelligent environments, removing computers from the desktop and making computing available anywhere and anytime. It must be said that the concept of intelligent environments is not new and began with home automation. The choice of name for the field varies somewhat from continent to continent in the English-speaking world. In general intelligent space is synonymous to intelligent environments or smart spaces of which smart homes is a subfield. In this collection, the terms *intelligent environments* and *ambient intelligence* are used interchangeably throughout. Such environments are made possible by permeating living spaces with intelligent technology that enhances quality of life. In particular, advances in technologies such as miniaturized sensors, advances in communication and networking technology including high-bandwidth wireless devices and the reduction in power consumption have made possible the concept of intelligent environments. Environments such as a home, an office, a shopping mall, and a travel port utilize data provided by users to adapt the environment to meet the user's needs and improve human-machine interactions. The user information is gathered either via wearable devices or by pervasive sensors or a combination of both. Intelligent environments brings together a number of research fields from computer science, such as artificial intelligence, computer vision, machine learning, and robotics as well as engineering and architecture. Other fields such as human-computer interaction and sociology deal with the human aspect of the problems.

May 2008
London and Tokyo

Dorothy Monekosso
Paolo Remagnino
Yoshinori Kuno

Contents

List of Contributors

Amedeo Cesta
Institute for Cognitive Science and Technology, National Research Council of Italy,
e-mail: amedeo.cesta@istc.cnr.it

Rita Cucchiara
Dipartimento di Ingegneria dell'Informazione, University of Modena and Reggio
Emilia, Modena, Italy,
e-mail: rita.cucchiara@unimore.it

Hatice Gunes
Faculty of Information Technology, University of Technology, Sydney (UTS),
e-mail: haticeg@it.uts.edu.au

Toshio Hori
Digital Human Research Center, National Institute of Advanced Industrial Science
and Technology, and CREST, Japan Japan Science and Technology Agency,
e-mail: t.hori@aist.go.jp

Luca Iocchi
Dipartimento di Informatica e Sistemistica, University of Rome, La Sapienza, Italy,
e-mail: iocchi@dis.uniroma1.it

Michie Kawashima
Faculty of Liberal Arts, Saitama University, Saitama, Japan,
e-mail: kawashima411@nifty.com

Yoshinori Kuno
Graduate School of Science and Engineering, Saitama University, Saitama, Japan,
e-mail: kuno@cv.ics.saitama-u.ac.jp

Riccardo Leone
Institute for Cognitive Science and Technology, National Research Council of Italy,
e-mail: riccardo.leone@istc.cnr.it

Dorothy N. Monekosso
Kingston University, Kingston upon Thames, London, UK,
e-mail: n.monekosso@kingston.ac.uk

Shin'ichi Murakami
Digital Human Research Center, National Institute of Advanced Industrial Science
and Technology and Graduate School of Engineering, Tokyo University of Science,
e-mail: s-murakami@aist.go.jp

Yasushi Nakauchi
University of Tsukuba, Tsukuba, Japan,
e-mail: nakauchi@iit.tsukuba.ac.jp

Daniele Nardi
Dipartimento di Informatica e Sistemistica, University of Rome, La Sapienza, Italy,
e-mail: nardi@dis.uniroma1.it

Yoshifumi Nishida
Digital Human Research Center, National Institute of Advanced Industrial Science
and Technology, and CREST, Japan Japan Science and Technology Agency,
e-mail: y.nishida@aist.go.jp

Federico Pecora
Institute for Cognitive Science and Technology, National Research Council of Italy,
e-mail: federico.pecora@istc.cnr.it

Massimo Piccardi
Faculty of Information Technology, University of Technology, Sydney (UTS),
e-mail: massimo@it.uts.edu.au

Andrea Prati
Dipartimento di Scienze e Metodi dell'Ingegneria, University of Modena and
Reggio Emilia, Modena, Italy,
e-mail: andrea.prati@unimore.it

Riccardo Rasconi
Institute for Cognitive Science and Technology, National Research Council of Italy,
e-mail: riccardo.rasconi@istc.cnr.it

Paolo Remagnino
Kingston University, Kingston upon Thames, London, UK,
e-mail: p.remagnino@kingston.ac.uk

S. Rush
Kingston University, Kingston upon Thames, London, UK,
e-mail: s.rush@kingston.ac.uk

Katsuhiko Sakaue
National Institute of Advanced Industrial Science and Technology, Tsukuba, Japan,
e-mail: k.sakaue@aist.go.jp

Yutaka Satoh
National Institute of Advanced Industrial Science and Technology, Tsukuba, Japan,
e-mail: yu.satou@aist.go.jp

S.A. Velastin
Kingston University, Kingston upon Thames, London, UK,
e-mail: s.velastin@kingston.ac.uk

Akiko Yamazaki
School of Systems Information Science, Future University-Hakodate, Hakodate, Japan,
e-mail: akikoy@fun.ac.jp

Keiichi Yamazaki
Faculty of Liberal Arts, Saitama University, Saitama, Japan,
e-mail: yamakei@post.saitama-u.ac.jp

Ikushi Yoda
National Institute of Advanced Industrial Science and Technology, Tsukuba, Japan,
e-mail: i-yoda@aist.go.jp

B. Zhan
Kingston University, Kingston upon Thames, London, UK,
e-mail: b.zhan@kingston.ac.uk

Chapter 1
Intelligent Environments: Methods, Algorithms and Applications

Dorothy N. Monekosso, Paolo Remagnino, and Yoshinori Kuno

1.1 Intelligent Environments

The roots of the research field can be traced back a few years starting with home automation. It has long been a dream to have a home that responds to the occupant's needs, anticipating needs and adapting to the occupant. The goals are generally to maximize comfort and safety, optimize energy usage, enhance general well-being, and eliminate strenuous repetitive activities. Research in this area takes on different labels in the wider research community around the world. Terms such as intelligent environment, smart spaces, smart homes, and ambient intelligence are often used interchangeably; not to mention related fields such as pervasive computing, ubiquitous computing, and wearable devices. Intelligent environments are living spaces with embedded sensors that sense and effectors that react to the occupants. The occupant need not wear or carry a computing-capable device, although wearable devices may augment and provide context awareness. The resurgence of the field is due in part to advances in technology that enable intelligent environments to be constructed. Particularly important are advances in sensor and actuator technology, and the development of networking including wireless technology as well as the miniaturization and the reduction in power consumption of the aforementioned devices and equipment. Miniaturization is critical as it not only allows devices to be embedded in the environment and out of sight, but also enables the devices to be made wearable.

Dorothy N. Monekosso
Kingston University, Kingston upon Thames, London, UK,
e-mail: n.monekosso@kingston.ac.uk

Paolo Remagnino
Kingston University, Kingston upon Thames, London, UK,
e-mail: p.remagnino@kingston.ac.uk

Yoshinori Kuno
Saitama University, Saitama, Japan, e-mail: kuno@cv.ics.saitama-u.ac.jp

D. Monekosso et al. (eds.), *Intelligent Environments*, Advanced Information
and Knowledge Processing, DOI: 10.1007/978-1-84800-346-0_1,
© Springer-Verlag London Limited 2009

1.1.1 What Is An Intelligent Environment?

It is often easier to describe with examples of applications than define. In this chapter, it is defined as a living or working space that "interacts in a natural way and adapts to the occupant". By natural it is meant in a manner that is natural for a human occupant and thus it implies speech. Adaptation refers to the fact that it learns to recognize and change itself depending on the identity and activity undertaken by the occupant with minimal intervention from the occupant. It must also adapt to natural changes such as those resulting from seasonal changes. In short, to qualify for the adjective intelligent, the environment must not only automate and facilitate everyday activities, it must also be adaptive and communicate in a natural way with humans. It must respond and adapt to users through learning. Early research targeted for the most part labor-intensive tasks in the home with the introduction of household appliances to reduce the impact of daily chores. These appliances were for the most part fixed at one location in the home. It is not uncommon nowadays to carry a number of smart portable devices such as PDAs and mobile phones but as with household appliances, their utility is limited as individual devices. The power of technology comes to light when all these can communicate and make collective decisions to enhance the user experience. More recently researchers, in the field of intelligent environments and related research fields have sought to extend the concept with intelligence everywhere and anywhere, connecting locations so that the intelligence travels with the user.

1.1.2 How Is An Intelligent Environment Built?

Intelligent environments are made possible by permeating spaces with intelligent technology that enhances quality of life. Environments range from private to public spaces and include the home, the office, shopping malls, airports and seaports, train stations to mention a few and utilize data provided by users to adapt the environment to meet the user's needs and improve human-machine interactions. The user information is gathered either via wearable devices or by pervasive sensors or by a combination of both. Thus the development of an intelligent environment requires a collaborative effort from a number of disciplines; from computer science, such as artificial intelligence, computer vision, machine learning, and robotics to mechanical and material engineering and architecture. Other fields such as human-computer interaction, psychology, and sociology deal with the human aspect of the problem.

1.2 Technology for Intelligent Environments

Advances in networking technology and the miniaturization of devices and sensors have made possible the intelligent environment. A typical home today contains a large number of embedded computers, each with a dedicated function and more often than not situated at a fixed location. Embedded computers are found not only in high-tech equipment such as PDAs, iPods, and mobile phones but also in the traditional household appliances such as cookers, washing machines, and fridges. Although useful as individual appliances and devices, their utility is greatly enhanced if connected through a network to allow communication between the devices and appliances. A system in which the fridge can command the main home computer to place an online order for milk because it is running low on milk is far more useful than any one of the appliances alone. In terms of sensing, of particular relevance to intelligent environments are technologies for visual and audio sensing. Intelligent environments will typically contain embedded cameras and microphones. There are a variety of other sensors of use in an intelligent environment, for example to monitor ambient characteristics such as temperature and humidity, to detect human presence through motion or pressure sensors, not forgetting bio-sensing for monitoring health status. In the next section, we shall see examples of these sensors deployed in private and public spaces. The sensors monitor the space and the humans within the space, allowing human activities to be recognized and behaviors analyzed. At the boundary between human and computer, interfaces are critical. The study of Human-Computer Interaction (HCI) grew alongside the developments in computer and related technology; greatly influencing technological design, for example display technology. Advances in display technology and the presentation of information therein have had a significant impact on the development of intelligent environment as a means of interaction. The applications of display technology can be seen in projects such as the iRoom at Stanford University. Similarly, information capture is no longer restricted to data entry using keypad entry systems. There are numerous technologies that facilitate information entry in more natural ways, from finger touch screens, handwriting and sketching recognition systems, to speech and gesture recognition systems. Some novel developments for capturing information are seen in the MIT Oxygen projects (e.g., AIRE). The goal in developing intelligent spaces is to enhance the user's experiences, maximize energy efficiency, and provide a safe and secure environment. The environment must adapt to one or more occupants and interact in a manner that is natural to humans. In order to adapt, the technology within the environment must sense and learn user preferences. Thus an important objective in intelligent environment research is to understand and predict human behavior and preferences based on sensed data be it audiovisual or other simple sensors such as temperature, and motion. It may be that the environment must understand the current activity in order to predict which effectors to activate next, e.g., which light to switch on or music to play; alternatively, it may be that the overall behavior is of interest as in monitoring health status for assisted living. The former can be achieved by modeling patterns in low-level actuator status while the latter can be achieved by modeling the occupant's patterns of behavior. Underlying technology that enables human behavior understanding and predictions includes detecting and tracking objects and person, persons and gesture recognition, gaze tracking from visual data, and speech and speaker recognition from audio data. Examples of these are discussed in the next section.

1.3 Research Projects

There are a number of projects in the area of intelligent environments taking place for the most part in universities and research centers around the world. We identify broad categories in intelligent environments research; the classification is based on the test bed or scenario. In the first category and probably the most extensively researched are private spaces which include homes. In the second category are public spaces such as meeting room, shared work environments or any other shared environment, e.g., elevator. The third category is the middleware that allows smart technologies to interact. In the next three sections, some examples are reviewed in each category. This is not a comprehensive list but a starting point; we do, however, attempt to review some of the most notable.

1.3.1 Private Spaces

Notable projects addressing problems of intelligent environments in the category of private spaces include the Intelligent Dormitory (iSpace) [2], the Adaptive Home project of Mozer [10], the MavHome project [5], and the Aware home [18]. The projects discussed tackle progressively more complex spaces and occupy larger test beds.

The **Intelligent Dormitory (iSpace)** [8, 9] is an ongoing project at Essex University (UK) aimed at developing an Intelligent Environment that continuously adapts so that the occupant does not explicitly interact with environment actuators. The goal is to make the environment economical, safer, and more comfortable. The test bed for the iSpace project is a furnished laboratory closely resembling a typical student room at the University of Essex. Technology within the iSpace monitors and learns the occupant's behavior. Sensors monitor while actuators allow the occupant to modify the iSpace environmental conditions. The devices include temperature sensors, humidity, motion sensors, and a matrix of light sensors across the room. The occupant's identity is determined by access control to the iSpace. In addition, a video camera allows external monitoring of the room. A number of actuators include: air circulators; fan heaters; a door lock actuator; motorized vertical blinds; automated window openers, and a light dimmer. The devices communicate with one another for coordinated action.

Among the first attempts at developing an adaptive home is Mozer [10] with the **Adaptive Home**. While iSpace focuses on a single room with a single occupant, the objective of the Adaptive Home was to *develop a home that essentially programs itself by observing the lifestyle and desires of the inhabitants, and learning to anticipate and accommodate their needs*. The goals are prediction to anticipate the needs of the occupant and efficient energy usage; for example, predicting the occupants' return to begin heating the home at an appropriate time and detecting statistical patterns in energy usage.

MavHome is a research project at the University of Texas at Arlington. The goal is to *create a fully integrated, versatile intelligent home that learns from its occupants and makes its own decisions for optimizing the home's operations. Foremost is maximizing the comfort, safety, security, health, money savings and enjoyment of the people who live there* [11, 12, 13]. MavHome takes the concepts found in iSpace and the Adaptive Home a step further. It is a completely integrated home and includes all the sensors for monitoring behavior and actuators for local and remote control found in the iSpace and the Adaptive Home. In addition, it continuously collects health data on occupants, alerting them to short-term and long-term changes in their health. Dealing with multiple occupants adds complexity to activity detection and behavior modeling. Other technology found in MavHome, include robots and smart appliances to aid occupants with reduced mobility. The refrigerator takes stock of inventory and replenishes itself by ordering groceries online, the microwave oven retrieves recipes online for dinner, while a robot vacuums the floors and another robot cuts the grass. The home entertainment system automatically records television programming it knows might be of interest to occupants. The project takes the concept even further to allow the networking of similar smart home for the purpose of energy conservation.

The Georgia Tech **Aware Home** [18] Research Initiative beginning in 1998 is a three-pronged project dealing with chronic care management in the home, future tools for the home, and digital entertainment and media. The test bed is at the Georgia Tech Broadband Institute Residential Laboratory, a purpose-built three-story, 5040-square-foot home for the design, development, and evaluation of technologies. In the chronic care management as well as monitoring as in other projects, interaction and visualization are key factors. A number of display tools have been developed for visualizing and displaying the results of health monitoring. The digital entertainment and media strand focuses very much on ease of use and networking of media devices as well as sharing media content within and between homes. Networking is a key feature in a truly intelligent environment; the future tools strand has for aim to simplify the setting up, security, and other networking related issues.

Little has been said so far about location-aware or context-aware application. An important issue in intelligent environment is localization. AwareHome addresses the issue with iCam and TrackSense. iCam simultaneously calculates its location and that of another object within the environment. TrackSense provides indoor positioning.

An important issue not often mentioned in the context of intelligent environment is user acceptance. The Aware project directly tackles acceptance issues as a related project.

1.3.2 Public Spaces

So far private spaces projects were described. Most if not all technology and tools developed in the context of private spaces are relevant to public spaces. The main differences encountered when dealing with public spaces relate to (a) the number of occupants, (b) different objectives (c) nature of activities and d) more complex privacy concerns. The number of occupants is potentially greater in public spaces such as airport, shopping mall and so detecting activities presents greater challenges. Although the overall goals of enhanced user experience, energy minimization, safety, and security are the same, the relative importance of the aims is different. For example, traveling through an airport, the aim is to get from the entrance to the plane with minimal stress and thus a key issue is personalized information presentation; tools and devices need to be location and context aware. The nature of activities in publics versus those in private spaces is more likely ephemeral and thus behaviors may be more difficult to model. Lastly but very important, privacy issues are more complex because acceptance of technology varies from person to person.

A number of research projects in the category of public spaces include MIT's Project Oxygen [19] and the Project AIRE, the Intelligent Classroom [21], the iRoom project [3], the HyperMedia Studio [4], and the Intelligent Elevator [22]. The physical space varies between the projects described, ranging from a single meeting room to larger halls and distributed areas. The description in this section focuses on tools and applications rather than devices. The latter are much the same as those described for private spaces.

The **Oxygen project** at MIT encompasses a group of projects. The Intelligent Room is a highly interactive environment that uses embedded computation to observe and participate in normal, everyday events, such as collaborative meetings. The Intelligent Room is populated with a number of tools purposely developed such as MeetingView, Annotea, and ASSIST. These augment a public space with basic perceptual sensing, speech recognition, and distributed agent logic. MeetingView is a tool to record the progress of a meeting in an intelligent meeting room capturing the format of the meeting and providing tools for analysis of the content.

Computer-based equipment and devices can be very useful tools but become increasingly difficult to operate or program with added functionality. D. Franklin at the University of Chicago has for goal to give devices and systems the capability to understand the user actions and respond so that the user can interact with the device in a manner that is more natural. In this project, the test bed is a classroom, the **Intelligent Classroom**. Cameras and microphones are employed to make sense of the speaker's actions and infer the speaker's intentions from those actions. The system attempts to assist the speaker by using knowledge of their intentions to decide on the next action; anticipating the speaker's actions. In the Intelligent Classroom, the speaker can concentrate on the lecture without learning to operate the classroom equipment; the Intelligent Classroom will assist when needed [20, 21].

The next project described differs somewhat from the previous in that the objectives are to investigate the design of rooms and integration of devices rather than creating an intelligent space for a specific purpose. Nevertheless the test bed is a public arena. The iRoom, an experimental research facility at Stanford University, is part of the **interactive workspaces (iWork)** project. The aim of the interactive workspaces project is to investigate the design and use of rooms containing one or more large displays with the ability to integrate portable devices and to create applications integrating the use of multiple devices in the space. The first project within the iWork larger project was the Interactive Mural, a four-projector tiled display. It included a pressure-sensitive floor, which tracked users in front of the display. A later iRoom contains three touch-sensitive whiteboard sized displays along the side wall, and a display with pen interaction called the interactive mural built into the front wall.

The **Hypermedia Studio** project brings together technology in the form of Intelligent Environments and arts. It creates original artistic works and systems combining interactive, performative, hypermedia content for both location-based applications (media/performance/installation events) and distribution-based applications (television, internet and dedicated networks).

In a class of its own but within the public space category is the intelligent elevator. Known as the **destination elevator**, this type of elevator is endowed with intelligent behavior. Destination-based elevators are so called because of the dispatching algorithm used that requires passengers to enter their destination floor on an entry device prior to entering the elevator car. Once the passengers enter the destination they are directed to an assigned car. Prior knowledge of the destination allows the control to dispatch in a most efficient manner. Once in the elevator, there is no more interaction between passenger and elevator control. The design of the data entry and displays is such that it is suitable for use by visually impaired or physically disabled passengers. Koehler and Ottiger [22] provide a review of the technology behind destination elevators.

1.3.3 Middleware

A requirement for an intelligent environment is the middleware that integrates devices, sensors, and effectors. Most projects have developed middleware that meets the requirements and integrates a specific system. An alternative approach is to develop generic reusable middleware. Microsoft in the context of their **EasyLiving** [1] project developed a prototype system for building intelligent environments. It is a software toolkit to construct intelligent environments. The prototype middleware comprised a distributed programming system of agent processes running on multiple computers, computer vision software to track people and maintain each person's identity, a database of a geometric model describing objects and people, and an event system and scripting system that can trigger actions based on people's movement.

While the EasyLiving middleware came from the development of an intelligent environment, the Amigo Project's main objective was to develop middleware. The aim was to develop middleware that dynamically integrates heterogeneous systems to achieve interoperability [14] between services and devices to connect in a network home appliances (heating systems, lighting systems, washing machines, refrigerators), multimedia and personal devices (mobile phones, PDAs). The network extends between homes of friends and relatives allowing the use of services across homes [15, 16, 17].

1.4 Chapter Themes in This Collection

This book comprises ten chapters by researchers from Europe, Asia, and Australasia and thus provides a broad view of the current state of Intelligent Environments on three continents. As discussed in the preface, the research field though now well established has a different designation in different parts of the world. Ambient Intelligence is a term well understood in Europe. Although the terms Intelligent Environments and Ambient Intelligence are not often used in Japan, a great deal of research has been conducted in Japan that directly relates to these topics. In particular, research has been carried out in pervasive computing and ubiquitous computing, as well in computer vision and robotics. We, the editors, have thus settled for a title that attempts to bridge over the linguistic constraints in the hope that the title adequately reflects the content for a global audience.

The Assisted Living (AL) theme runs through contributions from Hori et al. in Chapter 2 and Monekosso and Remagnino in Chapter 3. Quality of life is enhanced by creating an intelligent environment to support persons with reduced cognitive and/or physical capacity. Both address the issue of improving the care of the elderly with unobtrusive sensors embedded in the environment. Hori et al. address the problem in a communal environment, aimed at persons requiring intensive care, while Monekosso and Remagnino address the problem of caring remotely to allow independent living. More specifically, Hori et al. present an ultrasonic sensor system that detects and tracks wheelchairs and people. They show the applications of this system within the context of nursing care. In particular, the system is used to prevent users ("inhabitants" in their words) from falls and records the daily living activities of each user in order to provide better care. Monekosso and Remagnino present a system that can detect a deviation from an inhabitant's daily routine of activities and alert a caregiver or family. Still within the assistive living theme, Cesta et al. (Chapter 5) tackle the problem with a robot capable of supporting and assisting the inhabitant in their home. Nakauchi (Chapter 4) provides another example of assisted living though the target audience now is the wider population. In particular, Nakauchi presents a cooking-support robot system. The system learns the sequences of human actions from observing cooking and other human activities. It infers the next action from the learning results and suggests to the user what to do next through voice and gesture.

Key to many intelligent environments is the capability to identify individuals and recognize human behaviors within the environments. The most common sensor technology for this purpose is imaging. The imaging devices may be static as Yoda and Sakaue (Chapter 6) who propose a Ubiquitous Stereo Vision, a network of multiple stereo-vision systems that recognizes the actions of people in their environments. In addition to detailed observation of human activities in the laboratory, they describe several experiments in which people were monitored in open spaces, such as railroad crossings, train station platforms, and a pavilion at the Aichi Expo 2005. The theme is also found in Zhan et al. (Chapter 7).

A challenge for visual sensing is the field of view of the camera. Satoh and Sakaue (Chapter 8) propose a Stereo Omnidirectional System (SOS). The system is composed of twelve sets of three-camera stereo units (i.e., total 36 cameras) that provide omnidirectional color images and range data simultaneously in real time with a complete spherical field of view. As an application of the system they describe an intelligent wheelchair.

From imaging tools to applications in Ambient Intelligence, Prati and Cucchiara (Chapter 9) address visual sensing for very challenging applications; their environment is the urban outdoor. This has the additional complexity of numerous sources of data, and range of very interesting and non trivial applications. They address video data analysis, focusing in particular on the use of computer vision techniques for monitoring public parks.

As well as recognizing behavior it is often necessary to detect finer details in the environment such as the human affect. A major limitation of affective computing has been that most of the research on emotion recognition has focused on one single sensor modality at a time and especially the face display. Gunes and Piccardi (Chapter 10) introduce recent advances in multi-modal affect, focusing on systems that include vision as one of the input modalities, and attempt to analyze affective face and body movement either as a pure monomodal system or as part of a multi-modal affective framework.

Robots can be useful components in intelligent environments; to assist or support humans in need of assistance. Three chapters are related to robots. The first is the robot helper described by Cesta et al. (Chapter 5) that supports older persons in their home and the second is the cooking support system (Chapter 4) described by Nakauchi. Finally, Kuno et al. (Chapter 11) discuss the importance of vision in human-robot communication in the context of helper robots. They describe two cases: a helper robot and a museum guide robot. The helper robot can respond to simplified utterances with deixis or ellipsis by recognizing human actions with computer vision. The museum guide robot moves its head to display appropriate non-verbal information to human vision while explaining exhibits.

1.5 Conclusion

In this introductory chapter, intelligent environment was defined and the requirements for such an environment discussed. Research into intelligent environment evolved from the simple home automation projects to adaptive systems. With the proliferation of smart devices and networking in particular wireless networking, the future of intelligent environment research is integration of all technologies required to create a truly adaptive environment. More specifically, it is developing the necessary middleware that will allow devices from any sources to be plug and play and interact.

References

1. Brumitt, B., Meyers, B., Krumm, J., Hale, M., Harris, S., & Shafer, S.: EasyLiving: Technologies for Intelligent Environments. In: Proc. of the 2nd Int. Symp. on Handheld and Ubiquitous Computing, Lecture Notes in Computer Science, **1927**, 12–29 (2006).
2. The iDorm project home page: Intelligent Inhabited Environments Group, Department of Computer Science, University of Essex, Essex University, UK. Available via http://iieg.essex.ac.uk/idorm.htm. Cited 20/09/2007.
3. The iRoom project home page: Stanford Interactive Workspaces Project Overview (2007) Available via
   http://iwork.stanford.edu/. Cited 20/09/2007.
4. The HyperMedia studio project home page: UCLA HyperMedia Studio (2007) Available via http://hypermedia.ucla.edu/. Cited 20/09/2007.
5. The MavHome project home page: University of Texas, Arlington (2007) Available via http://cygnus.uta.edu/mavhome/. Cited 20/09/2007.
6. The Elite Care project home page: Elite Care Corporation, Milwaukie, OR, USA (2007) Available via
   http://www.elitecare.com/technology. Cited 20/09/2007.
7. Pollack, M. E.: Intelligent technology for an aging population: The use of AI to assist elders with cognitive impairment. AI Magazine **26(2)**, 9–24 (2005).
8. Doctor, F., Hagras, H.A., & Callaghan, V.: An Intelligent Fuzzy Agent Approach for Realising Ambient Intelligence in Intelligent Inhabited Environments. IEEE Trans. on Systems, Man and Cybernetics **35**, 55–65 (2004).
9. Rivera-Illingworth F., Callaghan. V., & Hagras H.A.: Neural Network Agent Based Approach to Activity Detection, in AmI Environments. In: IEE Int. Workshop on Intel. Environments, (2005).
10. Mozer, M. C.: Lessons from an Adaptive House. In: Smart environments: Technologies, protocols, and applications. ed. by D. Cook & R. Das. J. Wiley & Sons, 273–294 (2004).
11. Cook, D., & Das, S., Prediction Algorithms for Smart Environments. In: Smart environments: Technologies, protocols, and applications. ed. by D. Cook & R. Das. J. Wiley & Sons, (2004).
12. Das, S., & Cook, D.J.: Designing and Modeling Smart Environments. In: Int. Symposium on a World of Wireless, Mobile and Multimedia Networks (WoWMoM'06). 490–494 (2006).
13. Rao, S., & Cook, D. J.: Predicting Inhabitant Actions Using Action and Task Models with Application to Smart Homes. Int. J. of Artificial Intel. Tools **13(1)**, 81–100 (2004).
14. The home page of the Amigo project. Available via
    http://www.hitech-projects.com/euprojects/amigo/. Cited 20/10/07.
15. Vallée, M., Ramparany, F., & Vercouter, L.: Flexible composition of smart device services. In: The 2005 International Conference on Pervasive Systems and Computing (PSC-05), June 27-30, 2005, Las Vegas, Nevada, USA. (2005).
16. Kalaoja, J.: Analysis of vocabularies for Amigo home domain. In: Proceeding of 8th International Conference on Enterprise Information Systems, 23–27, May 2006 Paphos, Cyprus.
17. Vallée, M., Ramparany, F., & Vercouter, L.: Dynamic service composition in ambient intelligence environments: a multi-agent approach. In: Proceeding of the First European Young Researcher Workshop on Service-Oriented Computing, April 21-22, 2005, Leicester, UK.
18. The Aware Home project home page, 2005, University of Texas, Arlington Available via http://awarehome.imtc.gatech.edu/. Cited 20/10/07.
19. The Oxygen project home page Available via
    http://oxygen.csail.mit.edu/Overview.html. Cited 20/10/07.

20. The Intelligent Classroom project home page, Franklin, D., University of Chicago. Available via http://www.cs.northwestern.edu/ franklin/iClassroom/pubs.html. Cited 20/10/07.
21. Franklin, D., & Hammond, K.: The Intelligent Classroom: Providing Competent Assistance. In: Proceedings of International Conference on Autonomous Agents (Agents-2001).
22. Koehler, J., & Ottiger, D.: An AI-based approach to destination control in elevators. AI Magazine, Fall 2002.

Chapter 2
A Pervasive Sensor System for Evidence-Based Nursing Care Support

Toshio Hori, Yoshifumi Nishida, and Shin'ichi Murakami

Abstract This chapter introduces a pervasive sensor system for nursing homes, where the daily activities of inhabitants are continuously monitored. Deterioration in the quality of nursing care for inhabitants of nursing homes has become a major problem in an aging society. The authors address this problem with the use of pervasive sensors embedded in a nursing room. The sensors monitor the position of a subject and his wheelchair, then the information is utilized to provide prompt assistance to the subject and also to log their daily movement. In our experiments, we obtained the subject's position data for a month and a half in a nursing home in Tokyo and analyzed the subject's activity transitions, the number of times the subject went to a restroom, and other important factors for nursing care. This chapter presents the concept of evidence-based nursing care, overview of the current system and experimental results obtained.

2.1 Introduction

The rapid development of sensing and communication technology has enabled us to monitor daily activities in our living environment. It has also enabled intelligent spaces in which the behavior of inhabitants is monitored with pervasive sensors and services are provided based on the history and current conditions for the purpose of enhancing quality of life. This has become an important research field in computer

Toshio Hori

Digital Human Research Center(DHRC), National Institute of Advanced Industrial Science and Technology (AIST) and CREST, Japan Science and Technology Agency (JST),
e-mail: t.hori@aist.go.jp

Yoshifumi Nishida
DHRC, AIST and CREST, JST.

Shin'ichi Murakami
DHRC, AIST and Graduate School of Engineering, Tokyo University of Science.

D. Monekosso et al. (eds.), *Intelligent Environments*, Advanced Information and Knowledge Processing, DOI: 10.1007/978-1-84800-346-0_2,

science and robotics. When we develop such intelligent spaces, research on dealing with a variety of information obtained in our daily lives *quantitatively* must be emphasized especially because it is the key to provide useful and appropriate support for persons in the space. Indeed, our daily lives are the most familiar matter for us, but they have seldom been evaluated quantitatively. Therefore, we must pursue the methodologies of (1) monitoring persons and objects in the environment, (2) analyzing the data and abstracting meaningful information from the data, and (3) modeling and simulating daily lives of persons statistically.

Concerning the monitoring methodology, sensing technology in general and performance of sensors have advanced greatly but there are no or few principles for obtaining data of daily living. Even if there are a large number of sensors installed in a space, meaningful data cannot be obtained from these if the system designers do not have a clear view of the goal. That is, the monitoring methodology is the basis of the other methodologies and is influenced by them at the same time.

The data analysis and abstraction methodologies are closely related to the modeling and simulation methodology. When we develop a daily life model of a person and a simulator of his/her daily life, we must analyze the data obtained keeping in mind the objectives. That is, the model or the simulator regulates the analysis methods, and we cannot get any useful information or good models from the data if we use inappropriate analysis methods.

In this chapter, the authors introduce a pervasive sensor system for nursing homes. This is a collaborative research project with our research group and a nursing home in Tokyo, Japan. We visited the nursing home before developing the system and interviewed caregivers to investigate problems in the home. Then, we designed a sensor system to meet the requirements of caregivers and installed it in a nursing room in the home. This chapter presents the background of this collaborative research, system overviews and experimental results obtained in the nursing home.

2.2 Evidence-Based Nursing Care Support

2.2.1 Background of the Project

Statistics published in 2004 by The Ministry of Health, Labour and Welfare of Japan (MHLW) show that the number of caregivers working in nursing homes and that of old persons who use or stay in the homes were 62,306, and 49,7483, respectively [10]; this means that one caregiver has to support 8.0 old persons on average in the homes. The imbalance between the numbers of caregivers and old persons will likely increase and certainly impose a burden on caregivers, so we need to take some measures to alleviate the daily tasks of caregivers.

From our interview of caregivers in a nursing home, there are many tasks required of caregivers in nursing homes and they are classified into the following two types:

Direct tasks   Tasks in which caregivers have direct contact with inhabitants. They
                  include personal care, such as feeding, toileting, bathing, and sup-
                  port of transfer between a bed and a wheelchair.
Indirect tasks   Tasks in which caregivers do not contact inhabitants directly. They
                  include patrolling, recording daily reports of each inhabitant's ac-
                  tivities, and designing/evaluating the care plan of each inhabitant.

The direct tasks are mainly based on skills that can be instructed at nursing schools
and from skilled caregivers to novices. On the other hand, not all skills for the indirect tasks are transferable between caregivers.

For example, designing and evaluating care plans require precise and insightful
knowledge of each inhabitant, but such knowledge is not described *quantitatively*
but obtained through experience by each caregiver at present. That is, care planning and its evaluation rely principally on the intuition and experience of each caregiver, and the quality of nursing care and care plans varies between caregivers. To
overcome the situation and provide the same quality of services for all the inhabitants, *objective criteria* for designing/evaluating care plans are necessary in nursing
homes.

2.2.2 Concept of Evidence-Based Nursing Care Support

One of the candidates for the criteria mentioned above is the quantitative daily activity data of the inhabitants. If the activities of daily living (ADL) of each inhabitant are recorded quantitatively and presented to caregivers, they can understand
the condition of inhabitants much better than by relying only on their subjective
observations and intuition.

The system presented in this paper contributes to obtaining ADL of the inhabitants in the nursing home quantitatively by pervasive sensors. The authors call these
quantitative observation data the *evidence* (of ADL). If the system provides the same
evidence to all the caregivers, it will enable caregivers to establish effective care
plans and also to provide the same quality of services for all the inhabitants. The authors call this an evidence-based nursing care.

In the medical field, concepts of evidence-based medicine (EBM) and evidence-
based nursing (EBN) became popular in the late 1980's. EBM is the process of
systematically reviewing, appraising and using clinical research findings to aid the
delivery of optimum clinical care to patients, and EBN is the process by which
nurses make clinical decisions using the best available research evidence, their clinical expertise and patient preferences. Wallin et al. conducted research examining
the factors that promote EBN [19]. They concluded that supportive leadership, facil-
itative human resources, activity in seeking new research and enhanced implementation of research findings in clinical practice are the keys.

The evidence-based nursing care proposed by the authors is slightly different
from EBN. The word "evidence" in our concept indicates the quantitative ADL
data of each inhabitant obtained by the system whereas that word implies mainly

research evidence in EBN. That is, our concept concentrates on the personal evidence whereas EBN concentrates on the public evidence.

2.2.3 Initial Goal of the Project: Falls Prevention

When the authors started this collaborative research project with a nursing home in Tokyo, Japan, we were initially requested to develop a system to prevent falls/tumbles of the inhabitants using our ultrasonic sensor technology [8, 13].

According to the caregivers in the nursing home, wheelchair users tend to fall down from their beds or their wheelchairs when they transfer between bed and wheelchair or between wheelchair and toilet seat. It is also likely that old people who suffer from senile dementia tumble from their beds when they try to sneak out of the bed. This is statistically significant: a report from the MHLW in 2001 indicates that falls were the most common type of injury[1] (50.0%) that occurred in nursing homes in Japan and the third most common injury was tumbles (9.3%). That is, injuries in nursing homes will dramatically decrease if we can prevent falls and tumbles. This is one of the challenging research themes in welfare and this has been our initial goal of the project.

When we interviewed the caregivers, they told us that they had been using several kinds of sensors to detect and prevent such injuries from occurring. Figure 2.1 shows two types of sensors used in the home. Figure 2.1 (a) is a floor mat sensor at bed side, which detects falls from bed, and Figure 2.1 (b) is a touch sensor on a handrail of a bed, which detects a person sneaking out of the bed. However, those sensors had little effect on preventing injuries because (1) the inhabitants, though suffering from dementia, often learn how to escape being detected by the sensors, (2) the

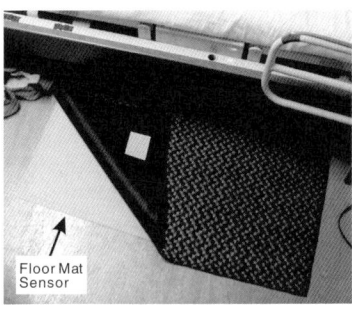 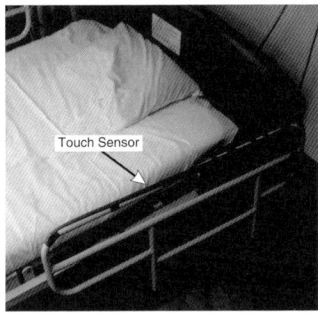

(a) Floor mat sensor at bed side   (b) Touch sensor on a handrail of a bed

Fig. 2.1 Sensors used in a nursing home

---

[1] In this paper, we do not use the word *accident* but *injury*, based on the concept that accidents do not occur by chance, but are preventable.

sensors often emit false alarms when, for example, a person turns over and touches a sensor by chance while sleeping, and (3) the sensors can *detect* but cannot *prevent* injuries even if they work correctly. Therefore, the fall prevention system requires the functions of (i) monitoring the subject all the time, (ii) detecting injury-prone activities beforehand, and (iii) calling caregivers in time to support transfer from/to a wheelchair before the fall injuries actually occur.

From the discussion with caregivers, we determined the following as the injury-prone activities which should be detected by the system: (a) entering a toilet, (b) approaching a bed, and (c) getting up and moving out of a bed, where transfer will occur after those activities. The system works as follows: it monitors the subject all the time and, when it observes one of those activities, it sends an alarm to caregivers by an existing nurse call line. As we had been developing an Ultra Badge system which obtains 3D position of small tags (we call the tag an Ultra Badge) continuously, we installed the system in one nursing room in the home.

But there was a problem. According to the caregivers, more than 90% of the inhabitants of the nursing home suffer from dementia so the caregivers cannot expect the inhabitants to put on any sensors by themselves. Moreover, we could not use contact-type sensors because they were often thrown away when the subject of our experiments felt uncomfortable. So we had to monitor the subject without attaching any sensors directly on his body. To solve this problem, we decided to attach a sensor (Ultra Badge) to a wheelchair which the subject uses. By monitoring the wheelchair's position, the system can detect activities (a) and (b). To detect activity (c), we developed a new sensor system. They are introduced in Sect. 2.4.

2.2.4 Second Goal of the Project: Obtaining ADL of Inhabitants

When the wheelchair locator subsystem began working stably in the nursing home, the authors had a meeting with the caregivers and showed them some experimental results of wheelchair tracking. Results shown were the trajectory of a wheelchair, such as Figure 2.11, during one-hour periods, and one of the results indicated that the subject went to the toilet one night, though no caregiver had noticed such activity at all.[2] Neither was there any written record of the activity. Watching the results, a caregiver indicated that the system was effective in obtaining ADL of inhabitants automatically.

As mentioned previously in Sect. 2.2.1, the number of caregivers in the nursing homes is much smaller than that of the old persons, so it is impossible to monitor activities of all the inhabitants all the time, especially during the night. On the other hand, caregivers are expected to design/evaluate care plans for each inhabitant, and require precise knowledge about each inhabitant for the care planning/evaluation. But currently, the planning task relies only on the intuition and experience of each

---

[2] We had not connected the system and a nurse call line yet because the system was unstable at that time.

caregiver. They need some means to obtain the conditions of each inhabitant properly and some objective criteria for designing/evaluating care plans.

The wheelchair locator subsystem monitors the position of a wheelchair only and it cannot know what the person is actually doing at all. But, according to the caregiver, even the position of each inhabitant is not always known to caregivers, as that experimental result revealed, so they wanted to know the location of the subject by the system. In that meeting, the caregivers requested the authors to analyze the position data as his ADL, and hence we set our second goal of this project as obtaining ADL of inhabitants.

Compared with the initial goal of the system, i.e., falls prevention, which is intended to be a prompt support for caregivers and inhabitants, the second goal aims at establishing long-term support for caregivers. The system obtains ADL of the inhabitants quantitatively and continuously as the evidence, and the caregivers provide proper nursing care for each inhabitant based on the evidence. As was mentioned before, this is the concept of evidence-based nursing care and, therefore, we call the system which supports this concept the Evidence-Based Nursing Care Support System.

2.3 Related Work

Stanford introduced a nursing home, the Oatfield Estates, in an article that he published in *IEEE Pervasive Computing* [18]. The home had employed IR (infrared) and RF (radio frequency) wireless communication tags not only for surveillance but also for monitoring the health condition of the inhabitants.

Sixsmith and Johnson [16] developed a fall detection system. They employed an array of IR sensors to obtain human posture images and detected falls by a neural network classifier. However, images were too coarse to detect falls correctly and the false recognition rate was high.

Srivastava et al. developed an embedded sensor network system for Smart Kindergarten [2, 17]. They used wearable small tags to capture interactions among students, teachers, and common classroom objects. They integrated many kinds of sensors, such as accelerometers, magnetic field, pressure, and light sensors, into the badge in combination with a DSP and a micro-controller, and the badge communicated with the environment-side system using Bluetooth.

Abowd and Price are co-directing The Aware Home Project [1, 11] at Georgia Institute of Technology. They built a three-story house as a living laboratory and embedded cameras, microphones, and other sensors in every room in the home. One of the focuses of their research is "Context-Awareness," that is, the system recognizes activities of the residents by embedded sensors and provides useful information for them based on the data obtained by the sensors.

The CareMedia is a research project at Carnegie Mellon University [7]. The researchers installed four cameras and microphones in a nursing home in Pittsburgh and recorded video images over a week. Analyzing the video, they tried to track the movement of each inhabitant and to extract their activities and interaction patterns at that place automatically [3].

Wilson solved the simultaneous tracking and activity recognition (STAR) problem [20]. He stated that people tracking can be improved by activity recognition and vice versa, then solved the problem using a particle filter.

Harada et al. developed small wireless devices for collecting life logs, which are the records of experiences in daily life [5]. They tried to predict behavior candidates which are likely to occur soon to provide appropriate supports [12].

There was an interesting study reported by Harmo et al. in the IEEE/RSJ IROS 2005 conference [6]. In this study, the authors surveyed the needs for elderly care by questionnaires and interviews given to the old people, caregivers and the general public. The paper suggested that we must consult the actual users about their needs when we develop real applications.

2.4 Overview and Implementations of the System

2.4.1 Overview of the Evidence-Based Nursing Care Support System

We employed ultrasonic sensors developed by the authors to remotely monitor the trajectory of a wheelchair and the position of the subject in bed. The system is composed of two subsystems: a wheelchair locator subsystem and an ultrasonic radar subsystem. Figure 2.2 shows a schematic diagram of the proposed system.

The wheelchair locator subsystem tracks the position of a wheelchair. We attached an ultrasonic emitter to the wheelchair of a subject and use this to track his position in a nursing room. In the room, there are a few areas defined where transfer

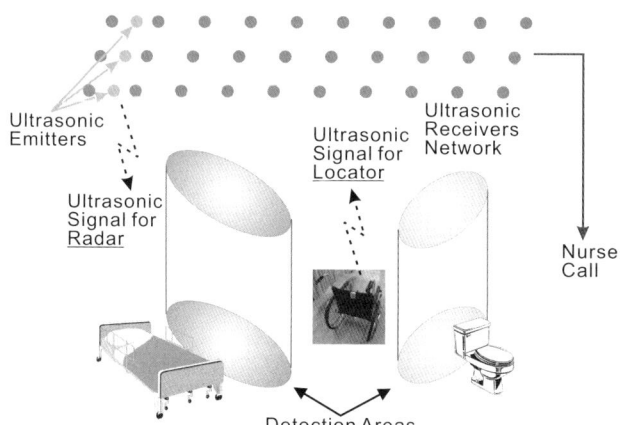

Fig. 2.2 Schematic diagram of the Evidence-Based Nursing Care Support System

from the wheelchair will occur. When the wheelchair enters one of those areas, the subsystem notifies caregivers of the occurrence of this "entering" event immediately by an existing nurse call system to support transfer [9].

On the other hand, the ultrasonic radar subsystem [15] monitors the position of a person on the bed. The subsystem uses time-of-flight of ultrasound pulses reflected by objects and monitors the position of the highest object on bed, on the assumption that the highest object in that particular area is the head or a part of the body of the subject. When it observes the highest object moving from the center of the bed to its edge, it assumes that the subject is getting out of bed and sends an alarm to caregivers just as the wheelchair locator subsystem does.

While the subsystems monitor the position of the wheelchair and the subject, the data obtained are stored in a log file as the ADL of the subject.

2.4.2 System Implementations

The ultrasonic 3D tag system developed by the authors [14] can track 3D positions of multiple ultrasonic tags continuously and concurrently by ultrasonic receivers embedded into the environment. We named the tag the *Ultra Badge*, and call the system the *Ultra Badge System.* We installed the system in a nursing room of the nursing home as the wheelchair locator subsystem.

Figure 2.3 shows two types of Ultra Badges developed by the authors. Figure 2.3 (a) is a small type whose size is approximately $3 \times 3 \times 1$ cm (in comparison with a Euro coin). It uses a button-type battery and works 6 to 8 hours. Figure 2.3 (b) is a long-battery-life type which uses a lithium-polymer battery and is active for one month. We use the latter type for this project because we must run the system continuously in the long term.

Figures 2.4 and 2.5 show the appearance of a nursing room and its floor plan, respectively, where we deployed ultrasonic receivers on the ceiling. Locations of ultrasonic receivers embedded on the ceiling are also displayed as intersection points of vertical and horizontal lines in Figure 2.5. The appearance of the ceiling and

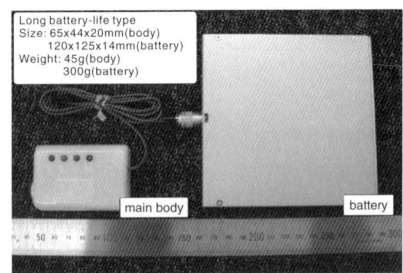

(a) Small type   (b) Long-battery-life type

Fig. 2.3 Ultra Badges—ultrasonic 3D tags developed by the authors—

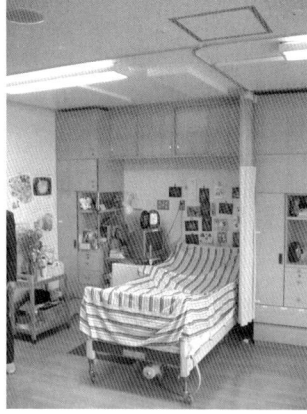

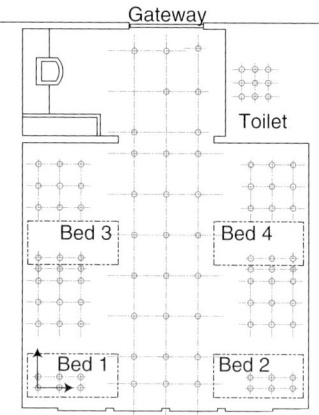

Fig. 2.4 Sensorized nursing room

Fig. 2.5 Floor plan of a nursing room and locations of ultrasonic receivers

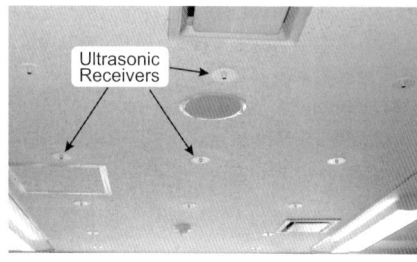

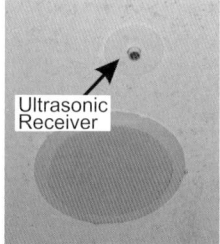

(a) Appearance of the ceiling   (b) Enlarged view

Fig. 2.6 Ultrasonic receivers on the ceiling of a nursing room

an enlarged view of an ultrasonic receiver are shown in Figures 2.6 (a) and (b), respectively. All the apparatus, such as receiver hubs and a data processing PC, are hidden inside the ceiling so the room looks the same as the other room except for the ceiling.

Figure 2.7 shows the configuration of the system from the viewpoint of apparatus and signal/data flows and Table 2.1 shows its specifications.

The system works as follows:

1. The controller PC sends the ID of an Ultra Badge to the SYNC signal generator.
2. The SYNC signal generator sends SYNC signal to all the receivers and the radio transmitter at the same time.
3. The radio transmitter broadcasts a Badge's ID as SYNC signal and a Badge which receives its own ID emits ultrasound pulses.
4. Concurrently, the receivers start their internal timers upon receiving SYNC signal to calculate the time-of-flight of ultrasound pulses.

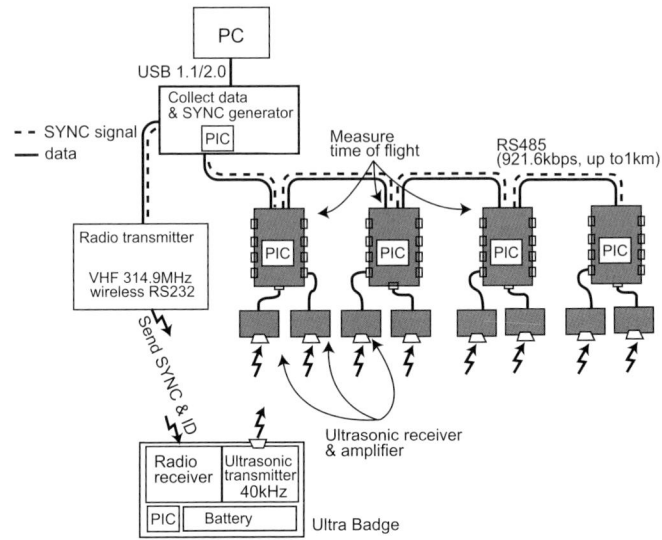

Fig. 2.7 Configuration of apparatus and signal/data flows of the Ultra Badge System

Table 2.1 Specifications of the Ultra Badge System

Position estimation error	less than 80 mm (50 mm average)
Resolution	15 mm (average)
Sampling frequency	up to 50 Hz
Frequency of ultrasound	40 kHz
A.P.L.* of ultrasound	51 dB
Badge size (main body)	65 × 44 × 20 mm
Battery life	About 1 month**

*A.P.L.: Acoustic Pressure Level.
**Using a lithium-polymer rechargeable battery.

5. Each receiver stops its timer upon receiving ultrasound pulses from the Badge and then the timer value is sent back to the PC from each receiver.
6. The PC collects all the receivers' timer values and computes Badge's position by triangulation using a robust estimator [4].

The current system monitors a single Badge attached to a wheelchair of a subject, but when there are several Badges in the environment, they are activated in turn.

Figure 2.8 shows the wheelchair and the Ultra Badge attached to the back of the seat. The battery box of the Badge is put inside a pocket at the back of the seat. As the subject cannot walk by himself but always uses the wheelchair to move around, the system assumes that the wheelchair's position indicates the subject's position.

Figure 2.9 shows an enlarged view of the ultrasonic radar subsystem. In this system, two ultrasonic emitters are coupled to control the intensity of acoustic pressure

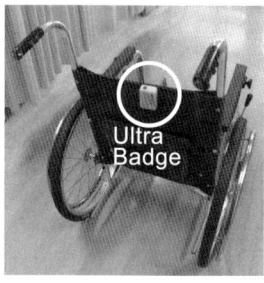

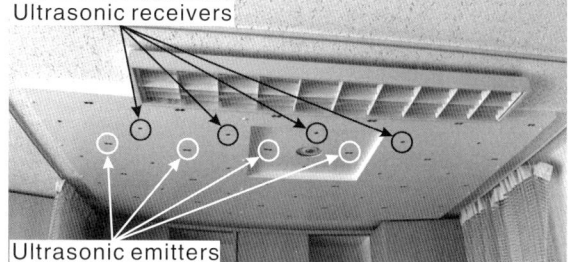

Fig. 2.8 A wheelchair to which an Ultra Badge is attached

Fig. 2.9 Ultrasonic radar subsystem

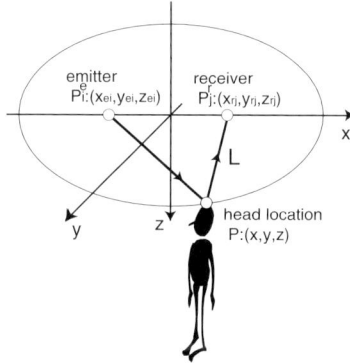

Fig. 2.10 Principle of localization of the ultrasonic radar subsystem

level and 24 pairs of emitters are embedded on the ceiling above the bed area. Ultrasonic receivers are shared with the wheelchair locator subsystem. Configuration of the system is almost the same as that of the wheelchair locator except for emitters which are connected to the SYNC generator by wire.

The principle used to locate the highest object by the radar subsystem is illustrated in Figure 2.10 and described as follows:

Let us suppose that the positions of the i-th emitter, the j-th receiver and the highest object (head, in Figure 2.10) are $\mathbf{P}_i^e$, $\mathbf{P}_j^r$, and $\mathbf{P}$, respectively, and the propagation distance is $L_{i,j}$, as shown in Figure 2.10. Then the following equation is obtained:

$$L_{i,j} = \|\mathbf{P}_i^e - \mathbf{P}\| + \|\mathbf{P}_j^r - \mathbf{P}\|. \tag{2.1}$$

This is an equation of a spheroid, where both $\mathbf{P}_i^e$ and $\mathbf{P}_j^r$ are its focal points and $\mathbf{P}$ is a point on its surface. As we installed many receivers on the ceiling, we can expect that more than three receivers receive reflected ultrasound pulses, and we obtain the same number of equations (2.1). When there are (at least) three spheroids whose axes do not coincide, we obtain the position of **P** as their intersection point, theoretically.[3] The subsystem activates the emitters in turn and computes the position of **P** every time. To minimize the ill effect of measurement errors, we use a robust estimator [4] which is the same as the wheelchair locator subsystem.

In both subsystems, ultrasound pulses sent through the air do not carry any personal information. Moreover, the system only collects distance between the Badge and each receiver calculated from time-of-flight data. he Badge's position is computed from distance data at a data processing PC. So personal information does not leak even if the ultrasound pulses are tapped. Therefore, we can say that the system respects and does not invade the privacy of the subjects.

2.5 Experiments and Analyses

Our proposed system was implemented in one room of a nursing home in Tokyo and its target was one old person (male) whose family allowed us to monitor and collect data related to his activities.

We attached an Ultra Badge to his wheelchair as shown in Figure 2.8 and recorded its position data for several months. The data was obtained at the sampling rate of 5 Hz by the system, and the position was estimated by a robust estimator [4] to remove the ill effect of outliers in the data set; consequently, their estimation error was less than 80 mm with an average of 50 mm. The data was sampled at 5 Hz and it was used for falls prevention (i.e., our initial goal), but such a high frequency is not required to know the ADL of the subject (i.e., our second goal) so the data was logged at 1 Hz.

The ultrasonic radar subsystem was so unstable during the experiment that we use only the data obtained by the wheelchair locator subsystem hereafter.

2.5.1 Tracking a Wheelchair for Falls Prevention

One of the purposes of the wheelchair locator subsystem was to track the wheelchair to prevent fall injuries. The subsystem monitored the wheelchair's position continuously, and sent an alarm to caregivers by an existing nurse call line when it detected the wheelchair entering into one of the areas where falls are likely to occur. Whenever the subsystem sent an alarm, it recorded the action into a log file at the same time.

We defined two areas, around the subject's bed and the toilet, as predefined areas, because transfer from/to wheelchair occurs in those particular areas. According to the log files, the system successfully detected the "entering" events and sent alarms

---

[3] Mathematically, three equations of spheroid give two solutions symmetrical to the plane on which an emitter and receivers exist. However, one of the solutions exists behind the ceiling, i.e., $z < 0$ in Figure 2.10, so we can easily omit it.

Fig. 2.11 Trajectory of a wheelchair obtained by the proposed system

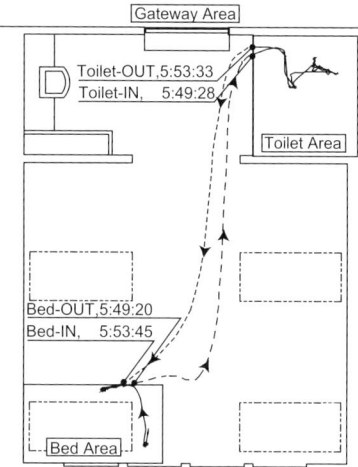

to caregivers. Therefore, we conclude that the subsystem was effective in predicting transfers.

Detecting the "entering" events gives binary information only, i.e., the wheelchair is *in* or *out* of the area. On the other hand, if we need more precise trajectory of the wheelchair in the room, it is available by plotting the data in the log files on the floor plan. For example, Figure 2.11 shows the trajectory obtained between 5 AM and 6 AM on one day. The data tells us that he went to the toilet once (around 5:50) and mostly stayed inside the bed area (probably on the bed) during the time period.

2.5.2 Activity Transition Diagram: Transition of Activities in One Day

Using the log recorded, we analyzed the ADL of the subject and developed his life log from the viewpoint of locations.

Table 2.2 shows the labels we used to describe the locations of the subject. To classify activities robustly, we set thresholds of elapsed time for each label. If this threshold is too long, such as 5 minutes, we cannot recognize short-term activities such as urination. On the other hand, if it is too short, such as 5 seconds, the position estimation errors fluctuate the classification when the wheelchair stays around the area boundaries. We empirically determined the thresholds as 30 seconds for robust classification.

For example, when the wheelchair stayed inside the Bed Area depicted in Figure 2.11 for more than (or equal to) 30 seconds, we assigned a label "Bed" for the time period. If the wheelchair got into the Gateway Area or the system failed to track the wheelchair for not less than 30 seconds, it was labeled as "Outside." When

Table 2.2 Labels and state values of activities from a viewpoint of locations

Activity	Location	Elapsed time	State Value
Outside	Gateway/outside*	≥ 30 sec	4
Toilet	Toilet area	≥ 30 sec	3
Bed	Bed area	≥ 30 sec	2
Moving	—	—	1

*The system cannot track wheelchair.

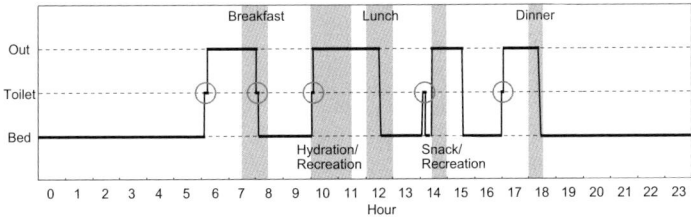

Fig. 2.12 Activity transition and common home schedule

the wheelchair was about the boundary of the bed area and its estimated position changed between inside and outside of the area frequently, a label "Moving" was assigned.

Using these labels, we made an activity transition diagram which describes the transitions of his activities each day.

Figure 2.12 illustrates the diagram plotted by the data obtained on one day (the label "Outside" is noted as "Out" in this figure). As it is difficult to recognize "Toilet" activities, they are circled. The activity "Moving" is not explicitly presented in the figure, but it is implicitly displayed as vertical links between activity states. In this figure, common home schedules are also shown as grayed rectangles. This figure tells us that (1) the subject went to the toilet in the room 5 times during the day, (2) he went out of the room 4 times, and (3) he spent his time mostly on his bed.

Our system is currently installed in one room only, and it cannot obtain the wheelchair's position at all when it goes out of the room. So a general label "Outside" is assigned to the time period while the wheelchair is outside. However, if the common home schedule is superimposed on this diagram as in the figure, we can assign more precise labels to each corresponding activity, such as breakfast, hydration and recreation.

2.5.3 Quantitative Evaluation of Daily Activities

The activity transition diagram provides a visual and comprehensible understanding of daily activities, so it is easy to recognize ADL of the subject by the diagrams.

However, it does not provide any quantitative evaluation of the living condition of the subject. Therefore, we calculated the correlation between activities of the subject and common home schedule to evaluate the condition quantitatively.

First, we assigned state values to the activities as in Table 2.2. Then, we made time series data of state transition for each experimental day and the common home schedule by using these values. As the activity raw data were logged at 1 Hz, the state values were assigned to the activity each second. As a result, we obtain two time series data; one is the activity state of the subject and another is the state of the common home schedule, both have 86,400 ($= 24 \times 60 \times 60$) data points.

Correlation coefficient, R, between the activity state and the common home schedule is calculated as follows:

$$R = \frac{\sum_t (S(t) - \overline{S})(S_C(t) - \overline{S_C})}{\sqrt{\sum_t (S(t) - \overline{S})}\sqrt{\sum_t (S_C(t) - \overline{S_C})}} \tag{2.2}$$

where t is time ($0 \leq t < 86,400$), $S(t)$ and $S_C(t)$ are the activity state value and the state value of the common schedule at time t, respectively, and $\overline{S}$ and $\overline{S_C}$ are their averages in one day.

Figure 2.13 shows 46 days of correlation coefficients calculated by equation (2.2). In general, we say that two data sets weakly correlate to each other when their correlation coefficient is between 0.4 and 0.7. The figure reveals that the coefficients of 33 days among 46 satisfy this criterion.

The correlation coefficient gives a quantitative understanding of daily activities of the subject. On the other hand, the activity transition diagram shown in Figure 2.12 gives a visual, quick and comprehensible understanding. When caregivers recognize that the coefficient value becomes lower, they can consult the diagram to find the cause and take measures if necessary. Therefore, they complement each other.

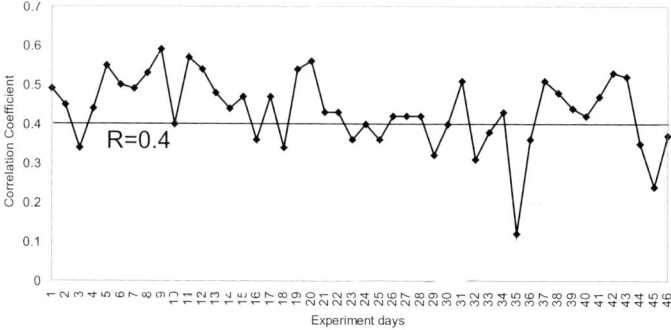

Fig. 2.13 Coefficient of correlation between the activity state and common home schedule

2.5.4 *Probability of "Toilet" Activity*

Falls are the most common type of injury in nursing homes in Japan as mentioned in Sect. 2.2.3, and one of the locations falls are apt to occur is at the toilet where transfer between a wheelchair and a toilet seat occurs every time. Ideally, each caregiver should pay close attention to all the inhabitants, detect toilet timing of each person, and support his/her transfer just in time, but this is not realistic. Currently, inhabitants who need transfer support are asked to call caregivers themselves using a nurse call line when they want to go to the toilet.

Therefore, if there is a system which can forewarn caregivers of the possible inhabitants who may go to the toilet within the next several minutes, caregivers can go directly to that particular person from a nurse station. As the first step toward developing such a system, we analyzed the probability of "Toilet" activity from the life log files.

First, we counted the number of activities of leaving bed (we call this "Move-FromBed" hereafter) and the number of "Toilet" activities immediately after "MoveFromBed." Then we obtained the probability of "Toilet" activity that follows "MoveFromBed."

Figure 2.14 shows the analysis results; the histogram indicates the numbers of "Toilet" and "MoveFromBed" activities and the line graph shows the transition rate, $P(t)$, where t is time. $P(t)$ is calculated as $P(t) = N_T(t)/N_B(t) \times 100(\%)$, by using the following notations:

$N_T(t)$ the number of "Toilet" right after "MoveFromBed" between t and $(t + 1)$ o'clock, and

$N_B(t)$ the number of "MoveFromBed" in the same time period.

From the figure, it is obvious that the subject got out of bed very frequently between 6 and 7 AM, but the probability of toileting during the time period was as

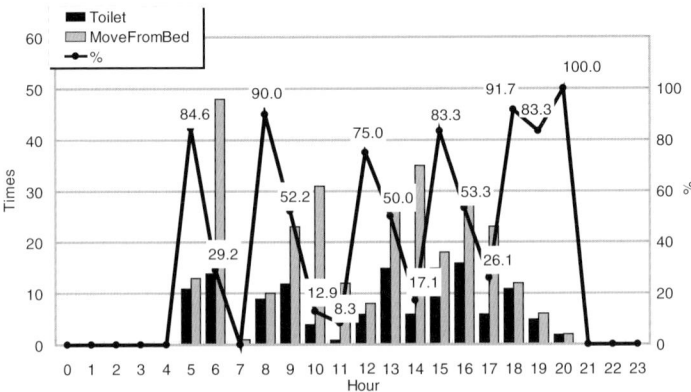

Fig. 2.14 Probability of "Toilet" activity

low as 29.2%. On the other hand, he did not get out of bed so often between 8 and 9 AM, but he went to the toilet with high probability of 90.0% at that time.

2.5.5 *Discussion of the Experimental Results*

We had run the wheelchair locator subsystem for several months. A controller program kept running during the experiments, but the tag sometimes did not work correctly when the battery ran out and caregivers did not notice it. Therefore, the system needs a notification function when to change and recharge the battery. Position estimation error was less than 80 mm and it was enough to obtain ADL of the subject.

The system needs to visualize the logged data to understand ADL of the subject, and one example of visualization is the trajectory of a wheelchair (Figure 2.11). When we showed caregivers several figures of the trajectory separated every one hour, their pro and con comments were as follows:

Pros (1) The resolution of the trajectory was more than they could imagine.
(2) If the system properly worked 24 hours a day, it would be very effective in obtaining ADL of the subject with little help from caregivers (see Sect. 2.2.4).
(3) To know such precise trajectory was effective in designing facilities such as width of corridors where two wheelchair users could pass easily and where to install handrails for the inhabitants.
Con Understanding the meaning of those figures was easy but the number of figures (24 figures/day) was too many. It was desirable to reduce the number to one or two figures per day.

Concerning the pros (3) comment, if we had employed other location sensors whose resolutions are coarser than ours, such as RFID tags whose resolution is about 30 to 50 cm, we could not obtain such precise trajectories. This is the point where our ultrasonic 3D tag system is superior to others in this application area.

Also, based on the con comment, we developed the activity transition diagram presented in Sect. 2.5.2 as another visualization method. The diagram provides a visual and comprehensible understanding of ADL of the subject.

The correlation coefficient is a quantitative representation of ADL and it provides information for a quantitative understanding of ADL of the subject. If we assume the conformity of ADL to the common home schedule as the evaluation criteria of the care plan, the correlation coefficient gives good information for the evaluation. On the other hand, the activity transition diagram is a symbolic representation of ADL and it gives visual, quick and comprehensible understanding. Taken together, the visual and quantitative measures complement each other.

Probability of "Toilet" activity revealed that there are several periods of time in a day during which the "Toilet" activities are likely to occur.

Concerning the activity transition diagram, the correlation coefficient of ADL, and the "Toilet" probability, caregivers' comments were as follows:

- The activity transition diagram and the correlation coefficient of ADL look very interesting because there are neither any visualization tools nor such precise activity data of the inhabitants at present. Current situation at the nursing home, i.e., a shortage of caregivers, does not allow caregivers to obtain activity data manually at such a high precision spatially and temporally, so the system is promising for supporting many tasks of caregivers.
- If the time when each inhabitant is likely to go to toilet is estimated, the caregivers can visit the inhabitant before he/she uses a nurse call and they can suggest toileting. This contributes not only to human resource allocation, but also to supporting the inhabitants who are apt to go to toilet alone.

For the latter comment, according to the caregivers, there is no explicit criterion for prioritizing multiple nurse calls which ring at the same time, so they cope with the calls in sequence in space (from the nearest room to the farthest) and/or in time (from the earliest call to the latest) currently. That is, sometimes several caregivers go to one inhabitant, while others are left unattended. If a system supports prioritizing multiple nurse calls and gives instructions on which call should be responded to first for each caregiver, it will be useful for efficient human resource allocation.

The caregivers of the nursing home that the authors have been collaborating with are very positive about deploying the system, but they do not have minute knowledge of the system, while the authors do not have full knowledge about what kind of information is useful and required for supporting nursing care. Thus, from our several years' experience, we learned that the communication between the developers and the users is the most important factor in developing a system and data processing/presentation methods which are of practical use in the real environment.

2.6 Prospect of the Evidence-Based Nursing Care Support System

The authors believe that what a nursing care support system should be is the evidence-based nursing care support system which enables users (i.e., the old persons) to select desirable services according to the level of serious illness or the degree of private information which they agree to disclose to the system.

To obtain the evidence, i.e., the quantitative activity logs, sensors for collecting ADL are the inevitable elements. However, the use of sensors must be considered carefully because they may easily invade privacy of the subject. This is a trade-off between the level of services the person can enjoy from the system and the degree of the privacy invasion. A nursing home must explain the pros and cons of the system to the prospective inhabitant and his/her family beforehand, and then obtain their agreement. In some cases, extensive use of privacy-invasive sensors, such as cameras, will be acceptable just as all kinds of medical apparatuses are allowed in ICUs (Intensive Care Units) of hospitals to save lives of patients.

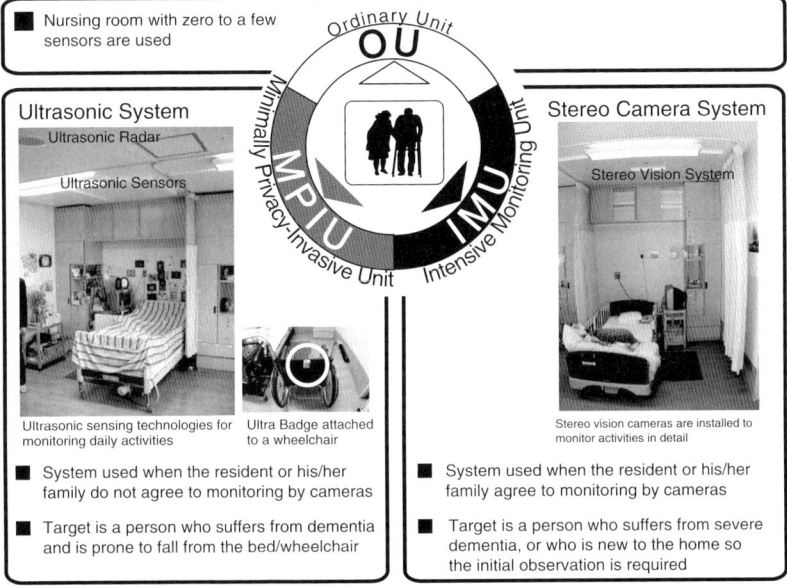

Fig. 2.15 Prospect of the Evidence-Based Nursing Care Support System

The authors' prospect of the Evidence-Based Nursing Care Support System is illustrated in Figure 2.15. Based on this prospect, we are planning to develop several types of nursing room in the nursing home:

1. **Intensive Monitoring Units (IMU)**—Rooms where any sensors, even cameras, are allowed to use. Equivalent of ICU in hospitals. Users of the room will be persons who suffer from severe dementia.
2. **Minimally Privacy Invasive Units (MPIU)**—Rooms where cameras are forbidden but other sensors which do not invade the privacy of the inhabitants are allowed. Possible users of this room will be persons whose level of dementia is not so severe but who are prone to fall from their bed/wheelchair, or who present wandering symptoms and sometimes sneak out of their bed during the night.
3. **Ordinary Units (OU)**—Rooms where zero to a few sensors are used. The inhabitants who do not need special observations will use this room.

The system presented in this paper is a kind of MPIU and we are now investigating a system which uses cameras to obtain activities on bed robustly.

2.7 Conclusions

This chapter introduced a pervasive sensor system for nursing homes. The authors investigated the problems in nursing homes, established a concept of evidence-based nursing care support, and clarified the functions required for the proposed system from the interviews with caregivers.

The authors developed a monitoring system by using ultrasonic 3D tag technology, installed it in a nursing room in a nursing home, and kept it running in real operation for several months. The wheelchair locator subsystem ran successfully and recorded the life log of the subject. The authors analyzed the log files and found his ADL patterns, such as the daily transition of ADL and the probability of "Toilet" activity. Then we summarized the experimental results with the pro and con comments of caregivers to claim the effectiveness of the system developed. Finally, we introduced our prospect of the evidence-based nursing care support system.

Acknowledgements The authors wish to express our great gratitude to Dr. Sachie Hasumura, Mr. Yuichi Motomura and Ms. Kazuyo Maruyama of the nursing home Aizenen for their great support in conducting this collaborative research project. We thank Mr. Hiroshi Aizawa, who was a master's course student with our research team, for his contribution to this project.

References

1. Aware Home Research Initiative: http://www.awarehome.gatech.edu/.
2. Chen, A., Muntz, R.R., Yuen, S., et al.: A support infrastructure for the smart kindergarten. In: IEEE Pervasive Computing, Vol. 1, No. 2, 49–57 (2002).
3. Chen, D., Yang, J., Wactlar, H.D.: Towards automatic analysis of social interaction patterns in a nursing home environment from video. In: 6th ACM SIGMM Int. Workshop on Multimedia Inf. Retrieval, in Proc. of ACM Multimedia 2004, 283–290 (2004).
4. Fischler, M.A., Bolles, R.C.: Random sample consensus: a paradigm for model fitting with applications to image analysis and automated cartography. In: Commun. of the ACM, Vol. 24, No. 6, 381–395 (1981).
5. Harada, T., Kawano, Y., Otani, S., et al.: Construction of wireless ad hoc network for lifelog based physical & informational support system. In: Proc. of the IEEE/RSJ Int. Conf. on Intell. Robots and Systems, 89–95 (2005).
6. Harmo, P., Taipalus, T., Knuuttila, J., et al.: Needs and solutions—home automation and service robots for the elderly and disabled. In: Proc. of the IEEE/RSJ Int. Conf. on Intell. Robots and Systems, 2721–2726 (2005).
7. Hauptmann, A.G., Gao, J., Yan, R., et al.: Automated analysis of nursing home observations. In: IEEE Pervasive Computing, Vol. 3, No. 2, 15–21 (2004).
8. Hori, T., Nishida, Y., Aizawa, H., Yamasaki, N.: Networked sensors for monitoring human behavior. In: Proc. of IEEE Int. Symp. on Computational Intell. in Robotics and Automation, 900–905 (2003).
9. Hori, T., Nishida, Y., Aizawa, et al.: Sensor network for supporting elderly care home. In: Proc. of the 3rd IEEE Int. Conf. on Sensors 2004, 575–578 (2004).
10. Japanese Ministry of Health, Labour and Welfare: Statistics on social welfare institutions (in Japanese), http://www.mhlw.go.jp/toukei/saikin/hw/fukushi/04/, (2004).
11. Kidd, C.D., Orr, R., Abowd, G.D., et al.: The Aware Home: A Living Laboratory for Ubiquitous Computing Research. In: Proc. of the 2nd Int. Workshop on Cooperative Buildings, 191–198 (1999).
12. Mori, T., Takada, A., Noguchi, H., et al.: Behavior prediction based on daily-life record database in distributed sensing space. In: Proc. of the IEEE/RSJ Int. Conf. on Intell. Robots and Systems, 1833–1839 (2005).

13. Nishida, Y., Aizawa, H., Hori, T., et al.: 3D ultrasonic tagging system for observing human activity. In: Proc. of the IEEE/RSJ Int. Conf. on Intell. Robots and Systems, 785–791 (2003).

14. Nishida, Y., Kitamura, K., Hori, et al.: Quick realization of function for detecting human activity events by ultrasonic 3D tag and stereo vision. In: Proc. 2nd IEEE Int. Conf. on Pervasive Computing and Commun., 43–54 (2004).

15. Nishida, Y., Murakami, S., Hori, T., Mizoguchi, H.: Minimally privacy-violative human location sensor by ultrasonic radar embedded on ceiling. In: Proc. of the 3rd IEEE Int. Conf. on Sensors 2004, 433–436 (2004).

16. Sixsmith, A., Johnson, N.: A smart sensor to detect the falls of the elderly. In: IEEE Pervasive Computing, Vol. 3, No. 2, pp. 42–47 (2004).

17. Srivastava, M.B., Muntz, R.R., Potkonjak, M.: Smart kindergarten: sensor-based wireless networks for smart developmental problem-solving environments. In: Proc. of the ACM SIG-MOBILE 7th Int. Conf. on Mobile Computing and Networking, 132–138 (2001).

18. Stanford, V.: Using pervasive computing to deliver elder care. In: IEEE Pervasive Computing, Vol. 1, No. 1, 10–13 (2002).

19. Wallin, L., Boström, A.M., Wikblad, K., Ewald, U.: Sustainability in changing clinical practice promotes evidence-based nursing care. In: J. Adv. Nurs., Vol. 41, No. 5, 509–518 (2003).

20. Wilson, D.H.: Assistive intelligent environments for automatic in-home health monitoring. Ph.D. Dissertation, Robotics Institute, Carnegie Mellon University, Pittsburgh (2005).

Chapter 3
Anomalous Behavior Detection: Supporting Independent Living

Dorothy N. Monekosso and Paolo Remagnino

Abstract In this chapter we describe a system for supporting independent living and enhancing quality of life of older persons. The home is equipped with non-intrusive standard home automation technology and an array of sensors that captures the status of appliances. Activities taking place in the home are detected and models of behavior for the occupant are created. In a first step, unsupervised classification techniques are employed to distinguish between activities that make up the occupant's daily routine. The activities include watching TV, entry to or exit from home, bathing, cooking, and eating. In a second step, a Hidden Markov Model technique is employed to model behaviors. In the context of assisted living, we define behavior as any detectable pattern in a sequence of activities. The models thus built can help a caregiver by distinguishing between normal and anomalous behavior. The system achieves this goal by predicting routine behavior. Behavior not recognized as routine is tagged as requiring investigation by a caregiver.

3.1 Introduction

This paper describes a system designed for the purpose of supporting independent living. The activities of the occupant are detected and a model of behavior built. In this context, a behavior is defined as a pattern in a sequence of activities. Activity is captured by an array of sensors embedded in the environment in such a way as to unobtrusively record daily activities. The objective is to discover patterns in the data leading to differentiation of activities and to discover patterns in sequences of activities described as behaviors. Thus models of activities and behavior are built. Predictive models can be used in a number of ways: to enhance user experience, to maximize resource usage efficiency (e.g., energy consumption),

Dorothy N. Monekosso and Paolo Remagnino
Kingston University, Kingston upon Thames, London, UK,
e-mail: n.monekosso@kingston.ac.uk, p.remagnino@kingston.ac.uk

D. Monekosso et al. (eds.), *Intelligent Environments*, Advanced Information and Knowledge Processing, DOI: 10.1007/978-1-84800-346-0_3,

to enhance safety and security. This work focuses on prolonging independent living and enhancing quality of life of older persons. To this end, the system must be capable of distinguishing between normal and anomalous activity. The steps are to detect activities, categorize these activities, detect trends and patterns in the activities and infer anomalous behaviors when these occur. The modeling of activities based on supervised learning techniques is described in [15]. The results indicate that supervised classification of activity is feasible. However, annotating raw data is time consuming. The work described here are the results of applying unsupervised techniques and behavior learning. With improved health care and living standards, an increasing number of people are living well into their old age. A challenge for society is to provide adequate care allowing older persons to maintain the desired level of autonomy while ensuring an enhanced quality of life. In a study conducted by Giuliani et al. [7] on the attitudes of older persons towards technology, the authors found against the stereotype of technology aversion by the elderly who would in fact be prepared to use it if benefits were perceived. In this study, a home was equipped with a number of sensors and actuators. The sensor and actuator status was recorded over a period of time and the resulting data used to model behavior. The sensors employed are small in size to facilitate embedding and no cameras are employed to promote acceptance by users. The relative simplicity of the sensor data output means that data preprocessing is relatively simple. However, the large number of sensors and the frequency of data capture result in a large data set which increases the complexity of data mining. In Section 3.2 published work related to the proposed system will be described. In Section 3.3 the methodology is described followed by a description of the experimental setup in Section 3.4. In Section 3.5 the results are presented and a discussion follows in Section 3.6 before concluding in Section 3.7.

3.2 Related work

In recent years, there have been significant advances in the field of intelligent environments. The advances were made possible in part by progress in sensor and device technology, network and computer technology. A number of research projects and a smaller number of fielded projects addressing issues of intelligent environments have been published. These include Microsoft's EasyLiving project ([4]), the Intelligent Dormitory iDorm [26], [6], the Interactive Room iRoom [27], the HyperMedia Studio [25], The MavHome project [5], [23], and the Evidence-Based Nursing Care Support System [9] to name a few and fielded applications such as The Elite Care project [24]. Technology is gradually gaining acceptance as a means to complement the work of caregivers and to assist persons with reduced physical or cognitive capacity in their day-to-day living. A review of published work, fielded systems and the state of the art in assistive technologies can be found in Pollack [18]. There are a number of ways in which an intelligent environment can be employed: to assist an individual in daily activities, to facilitate and complement the caregiver or to assess the person cared for. Recent advances in technology have made it possible to embed sensors into an environment. These range from arrays of relatively simple sensors that record on/off status for temperature, lighting ([14], [19]) to more complex sensors to record sound and images ([2], [3]). From these embedded sensors, various environmental attributes can be detected and activity inferred. The physical or cognitive status of the user/occupant can be inferred from the activity and a decision can be made regarding the health status of the occupant. A number of researchers are currently working on the problem of modeling human behavior based on input from multimodal sensor arrays. Supervised and unsupervised learning algorithms have been applied to learning a model of activity. In supervised model learning, [17] and [19] use naïve Bayes classification to identify activities. In the latter case, the data is supplied by a large number of very simple binary-valued sensors, while in the former case, the data source consists of PDA, keyboard, and telephone usage information. Brdiczka et al. [3] employ Bayesian classifier and Support Vector Machine to classify activities based on video data. An example of activity learning based on speech data is described by Brdiczka et al. [2]. A drawback of supervised learning is the need for a teacher to provide answers. This may take the form of annotated data. Typically the sensors will produce a very large amount of data that must be annotated. Unsupervised learning has the advantage of not requiring a teacher. Doctor et al. [6] model activity in the iDorm unsupervised employing fuzzy rule learning; Rivera-Illingworth et al. [22] employ neural networks. Mozer [14], in the ACHE project, employs reinforcement learning to learn models of behavior from observations for the purpose of predicting low level actuator status. In the MavHome project [5], [23], [21] predictive models are built based on Hidden Markov Model techniques. As with the ACHE project, the aim is to predict low-level activity such as on/off switching. By contrast, the work presented in this chapter aims to predict high-level activity and thus the occupant's behavior.

3.3 Methodology

The aim is to build a model to allow prediction of the next activity given the current activity and in so doing detect anomalous behavior. The first step is to differentiate between activities. The activities in the experiment are sleeping, bath (ablution), out/no activity, entry/exit, cooking, eating, working at a desk/study, and relaxation/watching TV. The second step is to discover patterns in the sequences of activities.

3.3.1 Unsupervised Classification Techniques

The techniques employed are based on clustering. This is the partitioning of the data set into subsets (clusters). The criterion for assigning a datum to a cluster is proximity according to some distance measure. Clustering algorithms are either hierarchical or partitional. In the former, clusters are found successively using previously found clusters. In the latter, all clusters are determined at once. Hierarchical algorithms are either agglomerative using a bottom-up approach or divisive using a top-down approach. Agglomerative algorithms begin with each datum as a separate cluster and recursively merge the clusters into larger clusters until the stopping criterion is satisfied. Divisive algorithms begin with the complete data set dividing it successively into subsets. The classification algorithms employed in the experiments are all well-established algorithms and include partitioning methods (KMeans, KMedoids, and EM - the latter is a probabilistic method) and a hierarchical agglomerative method. A brief overview of the similarities and differences [1] is presented here. Partitioning algorithms differ in the method by which the iterative relocation of points is performed. Methods such as KMeans [12] and KMedoids use a single point to represent a cluster. The KMeans centroid is the arithmetic mean of all points in the cluster. KMeans is thus sensitive to outliers. The KMedoid point is selected based on the location of a larger fraction of points and thus is less sensitive to outliers. KMeans is better suited to numeric data since the centroid is an arithmetic mean while the KMedoid can be used with nonnumeric data. Hierarchical algorithms use a proximity matrix for representing pairwise similarity. The similarity between two clusters can be determined as the minimum distance between elements of each cluster (single linkage). Alternatively the similarity is the maximum distance between elements of each cluster (complete linkage). Single linkage can cope with nonelliptical shapes but it is sensitive to noise and outliers. Complete linkage is less susceptible to noise and outliers because the similarity is determined by all pairs of points in the two clusters. However, it breaks down for large clusters [8]. In probabilistic partitioning methods, the cluster is identified with a model that consists of a mixture of distributions. The aim is to find the parameters of these distributions that maximize the log-likelihood. Advantages of the hierarchical methods over the partitioning methods are flexibility in terms of granularity and the use of any form of similarity or distance metric; however, the stopping criteria can be vague if it is not the number of clusters [1]. In partitioning methods, intermediate clusters are revisited for improvements while most hierarchical algorithms make no attempt at improving intermediate clusters. Partitioning methods generally suffer from time complexity. The distance measure used to determine the similarity between two points influences the shape of the clusters, as two points may be close according to one distance measure and far apart according to another distance measure. Some of the distance measures used are the Euclidean distance, the Manhattan distance, the Mahalanobis and the Normalized Google Distance (NGD).

3.3.2 Using HMM to Model Behavior

The Hidden Markov Model (HMM) is a statistical technique for modeling based on the assumption that the process is a Markov process with hidden parameters.

The states are not directly observable but the process has observable parameters and the hidden parameters can be determined. A Markov model is a stochastic state automaton in which a state has associated a prior probability and a set of transition probabilities. The prior probability of a process describes the probability of starting in a given state, and the transition probability describes the likelihood of a process moving into a new state. Evaluation of the HMM parameters requires calculating

- the prior probability π_i for each state S_i, representing the probability that a particular state is the starting state,
- the transition probabilities a_{ij} between two states S_i and S_j and
- the probability distribution function $b_j(O)$ of an observation vector O for a state, S_j, i.e., the conditional probability of a particular observation O given the state S_j.

Model selection can be performed by finding the model λ which yields the highest a posteriori likelihood $P(\lambda \mid O)$ given the sequence of N observations $O = (O_1, \ldots, O_N)$ associated with a time series. Reproducing the work of Rabiner [20], the probability of the observation set O given the model λ is

$$P(O \mid \lambda) = \sum_{q_1 q_T} \pi_{q_1} b_{q_1}(O_1) a_{q_1 q_2} \ldots a_{q_{T-1} q_T} b_{q_T}(O_T). \tag{3.1}$$

At each moment in time, the likelihood of a model given the current set of observations is calculated. The model λ that yields the highest a posteriori probability is the one currently providing the most likely interpretation, i.e., $\lambda^* = argmax_{\lambda \in a} P(O \mid \lambda)$.

The most likely model is calculated using the classical forward iterative procedure. The process is repeated until the termination stage in which the a posteriori probability of a model λ is computed by summing over all final values of the α variables computed for the model λ.

3.4 Experimental Setup and Data Collection

In this section, the experimental setup and the data collection are described. The array comprises temperature sensors, motion detectors, pressure mat, window and door status (open or closed), light level, light status (on, off and brightness level), smoke detector and Radio Frequency Identification Devices (RFID). Appliance status is captured by sensors. The nature of sensor output varied from binary valued data to continuous range. The sensors are located throughout a home, each room having at least one motion detector, one temperature sensor, one light level detector and two lighting status (on, off and light level setting) sensors. It is intended to use the RFID systems in future work as a means to disambiguate between users (occupant) of the home. RFID was not used in the experiment. In all data were captured by 47 sensors. For the purpose of the experiment, closed loop feedback control is disabled so that the inhabitant operates directly all actuators that control temperature, door/windows, and lighting. The control points (e.g., light switches) are located at the same positions as would be standard controls i.e., the actuator controls belonging to the experiment are located where one would expect to find them in a home so as not to skew the behavior and hence data collection and analysis. The reason for maintaining open-loop control is that users are monitored to allow a model of their behavior to be built. The monitoring took place over several periods each lasting a week; data logging was continuous, taking place night and day. The data were collected and analyzed using the algorithms described in Section 3.3.

3.4.1 Noisy Data: Sources of Error

There are various sources of error that add noise to the data collected. These are mainly due to intermittent failure of equipment to measure and/or record activities that trigger the sensors and are relatively rare. The sources of error are:

1. Failure of equipment to record an action, for example, a sensor failing to trigger. This was considered the most frequent and likely error.
2. Failure of equipment to take an action due to radio signal interference.
3. Random activation due to noise on mains or radio signal interference.
4. Sensor activation shorter than the sampling time during data processing. This happens particularly with motion detectors that trigger for a specific length of time.

The impact of these errors is discussed in this section. Type 2 error has no impact on data mining; the problem is user inconvenience as the action must be repeated, for example, the user must activate a switch a second time. Type 1 and 3 errors are problematic; the result is potentially one inaccurate datum (1) or missing datum (3). An example of this is a light recorded as on when it is in reality off or vice versa. These types of problems show up as an ON without an OFF i.e., a persisting ON/OFF beyond the expected duration. Graphical representations of the data very easily show outliers that can be further investigated. Note that errors in these two classes are relatively infrequent. Error type 4 results in missing data because the sensor ON time is shorter than the data processing sampling time. This problem is minimized by reducing the sampling interval. Heuristics can be used to perform some data cleansing; in addition, the heuristics can be inferred from data. These heuristics relate to temporal characteristics of sensor activity; examples are average ON/OFF times of sensors, exploiting the repetitive characteristics of activities.

3.4.2 Learning activities

The activities to distinguish are listed in Table 3.1 with the percentage instances in the data set. The nature of the different activities means there will always be an imbalance. The effect of this will be investigated in the future.

Table 3.1 List of activities and percentage of instances in the data set

Activity ID	Description	% instances
0	NO-ACTIVITY	50
1	BED TIME	3
2	ABLUTION	16
3	COOKING	6
4	EATING	6
5	RELAX-TV	1
6	ENTRANCE-EXIT	8
7	WORK	6

3.5 Experimental Results

In this section, the results are described. Unsupervised classifiers were employed. Prior to modeling, the data was processed, i.e., the raw sensor data converted to a vector time series. Each vector is of dimension n where n is the number of sensors.

3.5.1 Instance Class Annotation

The objective was to produce annotated sequences for the purpose of cross validation. Annotated dated is not used for modeling. Each sequence is a list of snapshots of the sensor status at discrete time steps. The sequences were then used to build a model of the activity over a period of time. Each sensor represents a dimension; there are 47 sensors. Sensors are either binary-valued or analogue with an output range of 0 to 1. The data were collected over a period of weeks. The raw data format is a time stamp followed by a sensor ID, name/description and status. A typical record is [15/02/2007 06:20:55 22855.7 0 BathMotionDetect 0]. During preprocessing, an additional attribute indicating period of the day was added. The day is divided into six periods. The raw data was annotated in a semiautomatic manner. Instances with two or more concurrent activities were annotated manually.

3.5.2 Data Preprocessing

Outliers can be identified and discarded if proved to be anomalous data resulting from equipment failure. Such data cleansing can be performed manually or automatically as described in this section. Manual data cleansing involved graphical representations to identify outliers. The automated method is a filtering process that makes use of two or more classification algorithms to generate an initial classification and remove instances that were misclassified (by all classifiers) from the data set prior to the modeling [29].

Identification of outliers Pre-processed data are displayed as histograms showing the frequency of occurrence of sensor active state. Examples are shown in Figures 3.1 and 3.2. The data were grouped into six classes corresponding to different times of the day. The time periods are 0–4, 4–8, 8–12, 12–16, 16–20, and 20–24. With some prior knowledge of the patterns of behavior of the occupant, outliers can easily be identified and removed from the data set as anomalous data points if explained by any one of the error sources described above. In Figure 3.1

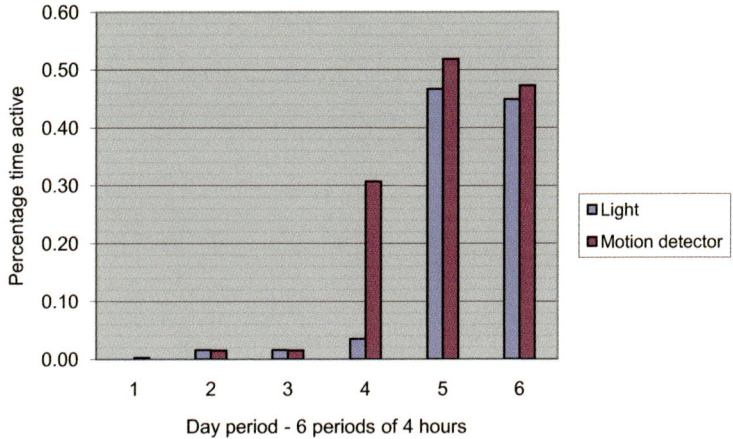

Fig. 3.1 Graphical representation of the active time (2 sensors) - no expected errors

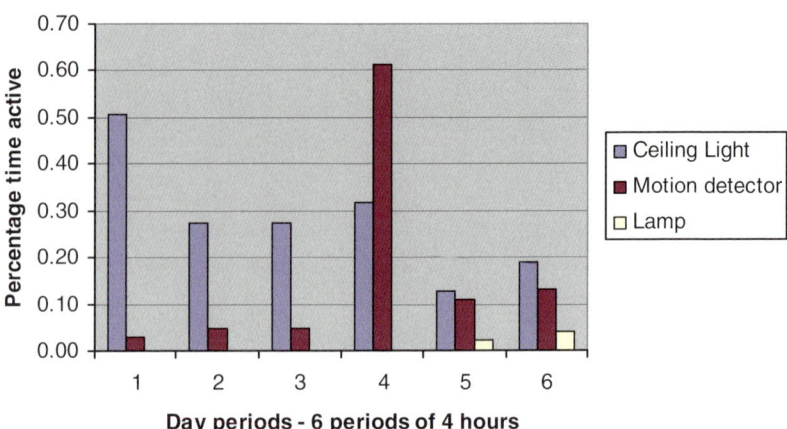

Fig. 3.2 Graphical representation of the active time (3 sensors) - outlier detected

given prior knowledge of the expected activities and most likely time of day for the activities, the histogram (trend) shows no outlier. By contrast, Figure 3.2 shows an outlier. It can be seen that a light sensor (purple, ceiling light) is active during all six periods. Investigation of the raw data indicated that the sensor status was incorrectly recorded; the light was active for an extended period of time during which it was known not to be active. One such outlier can significantly affect the overall trends. The overall modeling would be greatly biased by this outlier. There is functional redundancy in the system's design and therefore some errors may be corrected once outliers are identified. The above process can easily be automated to operate on a large data set.

Data filtering The removal of outliers can be automated employing a filter to remove misclassified instances prior to the final classification. In this method of cleansing, the filter consisted of a combination of classification algorithms; misclassified data were removed if they coincided in all [29]. In the experiments described here it was decided not to employ this technique as valid outliers can be removed. In general, removal of outliers by any means limits the capability of the system as an autonomous system.

3.5.3 Models: Unsupervised Classification: Clustering and Allocation of Activities to Clusters

The clustering algorithms were applied to the data and evaluated against the annotated classes (activities). Two experiments were conducted. In the first experiment clustering algorithms were applied to the processed data as discussed above. In the second, prior to applying the clustering algorithms, useless attributes were removed automatically. The selection criteria for removal were (1) nominal attributes having the same value for more than p percent of all examples and (2) numerical attributes with standard deviation less than or equal to a given deviation threshold. Applying these criteria to the data, 3 attributes were removed. KMeans, KMedoids and Agglomerative algorithms require the number of clusters as a parameter. This was set to 8 equal to the number of activities. Furthermore the Agglomerative algorithm employed the Euclidean distance measure and complete linkage. The EM clustering algorithm can determine the number of clusters through cross-validation; however, in this mode the performance was poor. In subsequent experiments the number of clusters was set to 8 for the EM algorithm. The results (classification accuracy) of cross-validated activities against the annotated data are shown in Table 3.2.

The Kappa statistic is a measure of the classification agreement that is not the result of chance. Based on Landis and Koch [11], a Kappa statistic in the range 0.41 to 0.6 indicates that the strength of agreement is moderate.

The results of cluster-activity association are shown in Table 3.3 . The association was performed using a supervised classifier (RIPPER). This classifier was selected because it performed well on the same data sets (training and test sets) in the

Table 3.2 Comparison of correct rates and kappa values

	Useless attributes removed		All used attributes	
	%Accuracy	Kappa	%Accuracy	Kappa
KMeans	73.4	0.659	76.5	0.595
KMedoids	66.73	0.601	74.32	0.601
EM	71.23	0.586	73.77	0.558
Agglomerative	65.90	0.559	71.23	0.457

Table 3.3 Comparison of activity classification sensitivity

	Class sensitivity (recall) (%)			
Activity	KMeans	KMedoids	EM	Agglomerative
A0	99.64	99.52	99.76	98.57
A1	98.70	98.70	98.70	97.40
A2	75.65	23.77	40.41	40.41
A3	0.00	76.41	0.00	18.03
A4	97.9	67.41	75.9	76.41
A5	0.00	0.00	0.00	0.00
A6	40.28	42.36	21.53	0.00
A7	0.00	0.00	98.00	52.00

supervised classification experiments [16]. In Table 3.3 the sensitivity is the relative number of examples correctly classified as positive among all positive examples. The class sensitivity is averaged over 20 runs. From Table 3.3, it is seen that there is not a one-to-one correspondence between clusters found and activities. All algorithms fail to generate a cluster corresponding to activity A5. The poor sensitivity for activity A5 may be explained by the relatively small percentage of instances for this activity in the training set (see Table 3.1). The number of examples for A5 represents only 1% of the total data set. According to Jain et al [10], the minimum amount of training examples required is at least 10 times the vector dimension. In this specific case this means approximately 6% of the total examples. All partitioning clustering algorithms with the exception of the EM algorithm fail to find a cluster that corresponds to activity A7 and the hierarchical algorithm performs poorly. The nature of the measurement system (sensors) meant that this activity is seen to overlap with other activities. The simple error rates (accuracy) in Table 3.2 would be adequate assuming that all errors have equal importance. In this application a false negative is more serious than a false positive. The latter represents a nuisance while the former could be life threatening. In addition certain activities may be considered to have greater diagnostic value in indicating the health status of the inhabitant.

Table 3.4 Confusion matrix for the Kmeans classifiers

	A0	A1	A2	A3	True class (%) A4	A5	A6	A7
A0	835	1	32	0	1	23	2	7
A1	2	76	10	28	1	8	3	0
A2	0	0	146	13	19	1	27	35
A3	0	0	0	0	0	0	50	0
A4	0	0	0	0	148	0	4	0
A5	0	0	0	0	0	0	0	0
A6	1	0	5	81	19	0	58	0
A7	0	0	0	0	0	0	0	0

For example, the activity Cooking is more informative watching TV. Another disadvantage of a simple rate is that it is dependent on the instance distribution. The confusion matrix provides more information by listing the true classification against the prediction for each class as shown in Table 3.4 for the KMeans classifier. In the case of the KMeans, looking along the diagonal, incorrect associations are made for two of the activities (A3 and A6) and no association for another two (A5 and A7). The impact of this is assessed when modeling behaviors.

3.5.4 Behaviors: Discovering Patterns in Activities

Learning patterns of normal behavior is a necessary step in detecting anomalous behavior, that is, detecting those behaviors deviating from the norm. Having discovered activities we now attempt to model the pattern in the sequence of clusters (activities). The behavior will be modeled as a sequence of activities. The model can then be used to predict the next behavior and so determine with a probability the occurrence of an anomalous behavior. A pattern of activities is modeled with a Hidden Markov Model. In this experiment, the behavior is modeled based on clusters identified in the unsupervised classification experiment. The observed states are the clusters. The number of symbols is set to the number of clusters. Models were thus obtained based on the classification produced by each of the techniques described in Section 3.3. Figure 3.3 shows the Log likelihood for each model as a function of number of hidden states. From the graphs in Figure 3.3, it is clear that irrespective of the number of hidden states selected, the model generated from the KMedoid cluster sequences produces the worst performance. The model based on the EM cluster sequence improves significantly with number of hidden states, surpassing the KMeans based model with 11 hidden states.

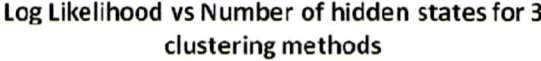

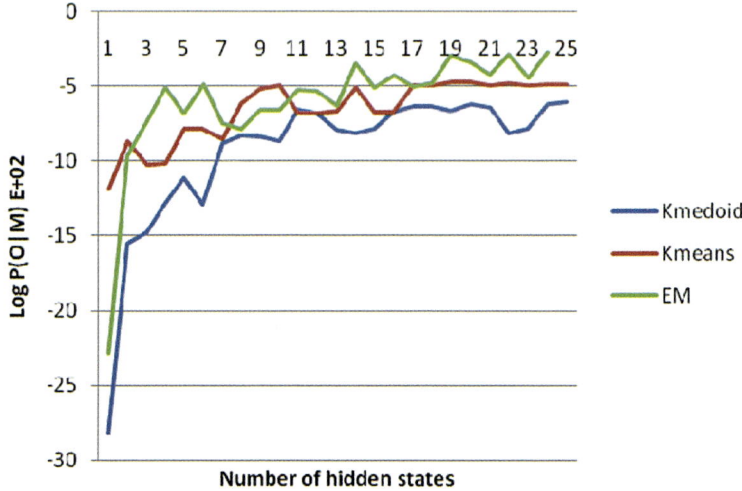

Fig. 3.3 Log likelihood as a function of hidden states (based on results of KMeans model)

The HMM model based on the KMeans clusters was selected for further assessment. Sequences of different length selected at random were generated from a test data set. The log likelihood as a function of hidden states for each of the sequence lengths chosen is shown in Figure 3.4. The graphs represent the average over 25 sequences for each length. From the graphs in Figure 3.4, it is clear that irrespective of the number of hidden states selected, the probability of the model generating the shorter sequences is higher.

3.5.5 Behaviors: Discovering Anomalous Patterns of Activity

The experiment described in this section was carried out to establish whether the system can detect anomalous patterns. To this end, a sequence length of 50 selected from the KMeans model (number of hidden states equal to 20) was chosen. Substitutions were introduced into this sequence, for example replacing cluster 7 with cluster 1 and obtaining the log likelihood for the new test sequence. This was repeated a number of times to obtain an average. The results are shown in Figure 3.5. The graph shows that in a sequence length of 50, the log likelihood is statistically unaffected for up to 10 substitutions and steadily worsens beyond that.

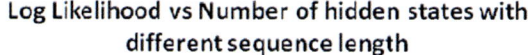

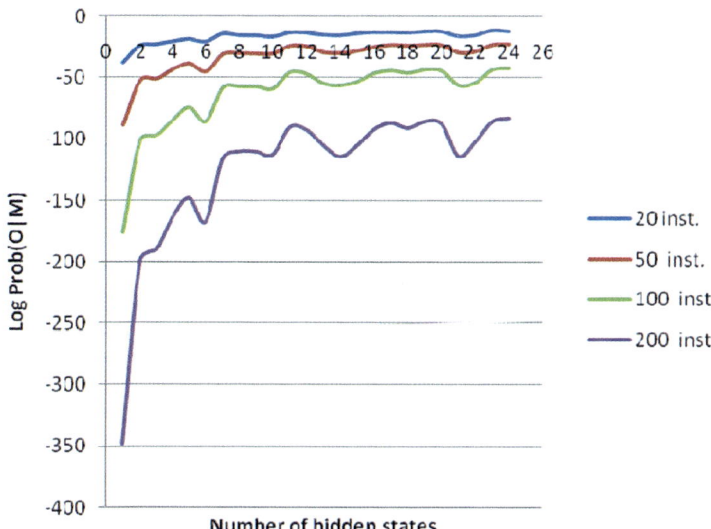

Fig. 3.4 Log likelihood as a function of hidden states for sequences of different length (inst. stands for instances in the sequence)

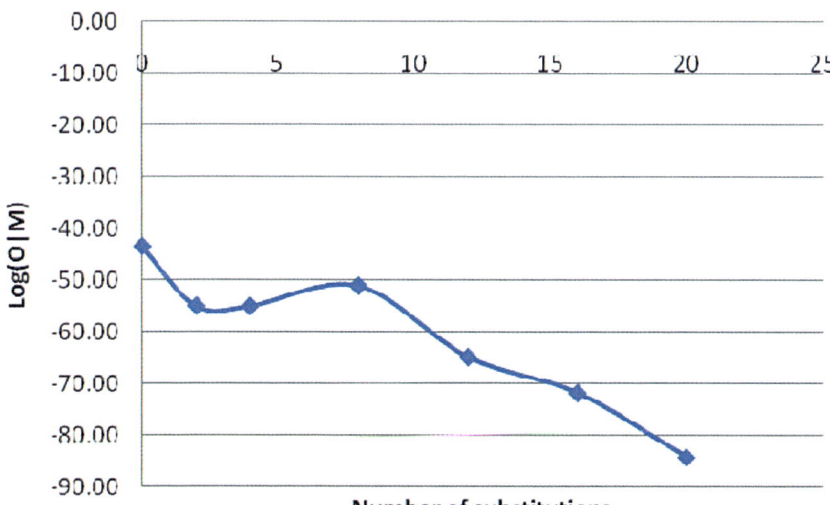

Fig. 3.5 Log likelihood as a function of cluster substitutions in a sequence

3.6 Discussion

A perceived drawback of learning in the context of an intelligent environment is
the necessity for the occupant to be trained to use the system at least for the duration of the (computational) model learning. This would be particularly unhelpful
in this context. In addition, as discussed in [14], there is a tendency for the user to
attempt to enforce regular habits which would skew the data. This work attempts to
overcome the problem by the use of very simple unobtrusive sensors. The selection
criterion for the sensors and actuators and placement of these within the environment is to mimic existing technology. For example, the light switches resemble and
operate in the same manner as ordinary light switches. A subset of the data set was
annotated for the purpose of cross-validating the results of unsupervised modeling.
The amount of data produced in a long-term investigation precludes the use of supervised classification. In addition, it is not always possible to observe inhabitants
for the purpose of collecting information for annotating sensor data. The results of
supervised classification have been described in previous work [16]; the unsupervised classifiers as expected performed less well. Validation of the unsupervised
classifier results against annotated data indicates activity detection rates up to 76%.
Experimental results showed that any filtering produced only marginal differences
and did not necessarily improve the overall results. The cluster-activity association
is not necessary in order to detect anomalous behavior. Models of behavior were obtained using an HMM algorithm with the clusters as observable states. The HMM
generatedmodels were assessed by plotting the average probability of a sequence
of observations given the model for various lengths of sequence. The capability
of the system to detect anomalous behavior was assessed by investigating the effect of cluster substitutions in sequences produced by the KMeans classifier. The
prediction for anomalous behavior is given in term of log-probability; comparing
an incoming new sequence to sequences representative of normal behavior. In real
terms this means how probable that the lack of kitchen activity at a given time represents a condition for raising the alarm. The decision to act or not to act remains
with the caregiver. Additional information regarding past activities within a time
window may be provided to help the decision making. The action of the caregiver
may also be graded according to the probability assigned to the anomalous behavior.
For example if the alarm is raised with a probability below a predefined threshold,
then the caregiver may choose in the first instance to provide remote assistance. It is
not considered a requirement to achieve 100% detection rates for the system to be
regarded as successful. However, failure to detect an anomalous behavior will result
in either false alarms or in missed alarms. The former is not considered a major issue
if the false alarm rate is low. However the latter may present a danger in the context of care in terms of delayed warning. The system performs better with mutually
exclusive activities, that is, only one activity can take place at any given time. This
assumption will not always hold true; particularly if there is more than one occupant. Multiple occupancy can be dealt with by other types of sensors, e.g., voice
recognition or RFID technology worn by the inhabitants. The issue of concurrent
overlapping activities has been addressed in the context of multiple occupancy. The

method relies on the fact that one important characteristic of multiple occupancy is concurrent activities. Multiple activities are classified but the different occupants are not distinguished.

3.7 Conclusions

In previous work, supervised classification was investigated, but annotating data is a long process and would be difficult as the number of sensors and number of activities increased. In some circumstances, it may not be possible to observe inhabitants in order to gain sufficient information for annotation. In this paper, the results of unsupervised classification of activities based on data gathered from multiple simple sensors installed in a home are described. The aim was to identify activities based on information such as lights status, motion detectors, pressure mat status, and create a model of daily activity that will help in detecting anomalous behavior. The results indicate that unsupervised classification is inferior; however, for anomalous behavior detection based on HMM modeling activity classification is not a prerequisite. The HMM technique was employed for modeling behavior as a sequence of activities. Future work is aimed at improving the performance of unsupervised classification and carrying out a formal assessment of the system's ability to detect anomalous behavior over an extended period of time is needed.

References

1. Berkhin, P.: Survey of Clustering Data Mining Techniques. Accrue Software (2002) Available via.
   http://www.ee.ucr.edu/~barth/EE242/clustering-survey.pdf.
   Cited 10/07/2002
2. Brdiczka, O., Vaufreydaz, D., Maisonnasse, J., & Reignier, P.: Unsupervised Segmentation of Meeting Configurations and Activities using Speech Activity Detection. In: Proc. of the 3rd IFIP Conf. on Artificial Intell. Applications and Innovations, 195–203 (2006)
3. Brdiczka, O., Reignier, P. & Crowley, J.: Detecting Individual Activities from Video in a Smart Home. In: Proc. of the 11th Int. Conf. on Artificial Intell. Knowledge-Based and Intelligent Information and Engineering Systems, 195–203 (2007)
4. Brumitt, B., Meyers, B., Krumm, J., Hale, M., Harris, S., & Shafer, S.: EasyLiving: Technologies for Intelligent Environments. In: Proc. of the 2nd Int. Symp. on Handheld and Ubiquitous Computing, Lecture Notes In Computer Science; **1927**, 12–29 (2006)
5. Cook, D., & Das, S., in *Prediction Algorithms for Smart Environments*, ed. by D. Cook & R. Das. by Smart Environments: Technology, Protocols and Applications. J. Wiley & Sons, 175–192 (2004)
6. Doctor F, Hagras H, Callaghan V.: An Intelligent Fuzzy Agent Approach for Realising Ambient Intelligence in Intelligent Inhabited Environments. IEEE Tran. on Systems, Man and Cybernetics **35**, 55–65 (2004)
7. Giuliani, M. V., Scopelliti, M. & Fornara, F.: Elderly People at Home: Technological Help in Everyday Activities. In: Proc. of the 14th IEEE Int. Conf. on Robot and Human Interactive Communication, 365–370 (2006)

8. Grossman, R.L., Kamath, C., Kegelmeyer, P., Kumar, V. & Namburu, R.: Data Mining for Scientific and Engineering Applications. Kluwer Academic Publishers (2001)
9. Hori, T., Nishida, Y., & Murakami, S., in *A Pervasive Sensor System for Evidence-based Nursing Care Support*, ed. by D. Monekosso et al, In this volume, Springer (2008)
10. Jain, A. K., Duin, R. P. W. & Mao, J.: Statistical pattern recognition: A review. IEEE Tran. on Pattern Analysis and Machine Intelligence **22(1)**, 4–37 (2000)
11. Landis, J.R. & Koch, G.G.: An Application of Hierarchical Kappa-type Statistics in the Assessment of Majority Agreement among Multiple Observers. Biometrics. **33(2)**, 363–374 (1977)
12. MacQueen, J. B.: Some Methods for classification and Analysis of Multivariate Observations. In: Proc. of the 5th Berkeley Sym. on Mathematical Statistics and Probability, University of California Press, **1**, 281–297
13. Mitchell, T., : Machine Learning. McGraw-Hill (1997)
14. Mozer, M. C., in *Lessons from an adaptive house*, ed. by D. Cook & R. Das. Smart environments: Technologies, protocols, and applications, 273–294 (J. Wiley & Sons, 2004)
15. Monekosso, D. N.: Modelling Behaviour: Supporting Independent Living. In: Proc. of The European Conference on Ambient Intelligence, Workshop on Assisted Living, to appear, (2007).
16. Monekosso, D. N., and Remagnino, P.: Monitoring Behavior with an Array of Sensors, Computational Intelligence, to appear, 2008.
17. Mühlenbrock, M., Brdiczka, O., Snowdon, D., and Meunier, J.-L.: Learning to detect user activity and availability from a variety of sensor data. In: Proc. of the 2nd IEEE Conf. on Pervasive Computing and Communications, **14(17)**, 13–22 (2006).
18. Pollack, M. E.: Intelligent technology for an aging population: The use of AI to assist elders with cognitive impairment. AI Magazine. **26(2)**, 9–24 (2005)
19. Tapia, E. M., Intille, S. S. and Larson, K.: Activity recognition in the home setting using simple and ubiquitous sensors. In: Proc. of PERVASIVE, LNCS **3001**, Ed. A. Ferscha & F. Mattern, Springer-Verlag, Berlin Heidelberg, 158–175 (2004).
20. Rabiner, L.R.: A tutorial on HMM and selected applications in speech recognition. Proc. of the IEEE. **77(2)**, 257–286 (1989).
21. Rao S. and Cook, D. J.: Predicting Inhabitant Actions Using Action and Task Models with Application to Smart Homes. Int. J. of Artificial Intel. Tools **13(1)**, 81–100 (2004).
22. Rivera-Illingworth F., Callaghan V, & Hagras H.A.: Neural Network Agent Based Approach to Activity Detection, in AmI Environments. In: IEE Int. Workshop on Intel. Environments, v2–92, (2005).
23. Das, S., Cook, D.J.: Designing and Modeling Smart Environments. In: Int. Symposium on a World of Wireless, Mobile and Multimedia Networks (WoWMoM'06). 490–494 (2006).
24. The Elite Care project home page: Elite Care Corporation, Milwaukie, OR, USA (2007) Available via.
`http://http://www.elitecare.com/technology.Cited20/09/2007`
25. The HyperMedia studio project home page: UCLA HyperMedia Studio (2007) Available via.
`http://hypermedia.ucla.edu/.Cited20/09/2007`
26. The iDorm project home page: Intelligent Inhabited Environments Group, Department of Computer Science, University of Essex, Essex University, UK (2007) Available via.
`http://iieg.essex.ac.uk/idorm.htm.Cited20/09/2007`
27. The iRoom project home page: Stanford Interactive Workspaces Project Overview (2007) Available via.
`http://iwork.stanford.edu/.Cited20/09/2007`
28. The MavHome project home page: University of Texas, Arlington (2007) Available via.
`ttp://cygnus.uta.edu/mavhome/.Cited20/09/2007`
29. Witten, I.H. and Frank, E.: Data Mining: Practical Machine Learning Tools and Techniques. 2nd edn. (Morgan Kaufmann, 2005).

Chapter 4
Sequential Pattern Mining for Cooking-Support Robot

Yasushi Nakauchi

Abstract Recent technological advances have meant that many electrical household appliances are computer-controlled and can be networked. In this chapter, we introduce a human activity recognition system, which infers the next human action by taking account of the past human behaviors observed so far. Based on the recognition system, we have developed a cooking support system which uses an LCD touch panel on the kitchen counter and a mobile robot on the floor. The system suggests the next action to carry out using voice and gestures. Experimental results confirm the feasibility of the inference system and the quality of the support is investigated.

4.1 Introduction

Recent technological advances have meant that electrical household appliances carry a processor and can be networked. If the environments within which we live can recognize our activities indirectly through sensors, novel services to support our activities can be developed. This idea was first proposed by Weiser as **ubiquitous computing** [18] and has emerged as Aware Home [5], Intelligent Space [9], Robotic Room I, II [17, 12], Easy Living [3, 8], Smart Rooms [15, 16], etc.

One of the most important factors for such systems is the recognition of human behavior using ubiquitous sensors. In the Intelligent Space system, the position of a human is detected with the use of multiple cameras on the ceiling allowing the human to be followed by a mobile robot [9]. Easy Living also detects the position of a human and switches on a light in close proximity to the human [3, 8]. These systems provide services by taking into account human intentions where the intention is inferred from the location of the human. On the other hand, one of the applications in Robotic Room I requires the human to express explicitly his intention. Robotic

Yasushi Nakauchi
University of Tsukuba, Tsukuba, Japan, e-mail: nakauchi@iit.tsukuba.ac.jp

D. Monekosso et al. (eds.), *Intelligent Environments*, Advanced Information
and Knowledge Processing, DOI: 10.1007/978-1-84800-346-0_4,

Room I will recognize a patient lying on a bed pointing at an object using the vision system and the robotic manipulator will hand the object to the patient [17].

In order to recognize implicit human intentions, Asaki et al. have proposed a human behavior (i.e., changing clothes, preparing meals, etc.) recognition system based on state transition models [1]. Moore et al. have proposed a Bayesian classification method, which enables recognition of the various kinds of human behavior through a learning mechanism [10, 11]. We also developed a ubiquitous sensor room, Vivid Room (see Figure 4.1), and proposed a human intention recognition system employing an ID4-based learning algorithm [13]. With the proposed learning algorithm, we succeeded in recognizing what a human intends to do such as studying, eating, arranging, and resting (see Figure 4.2).

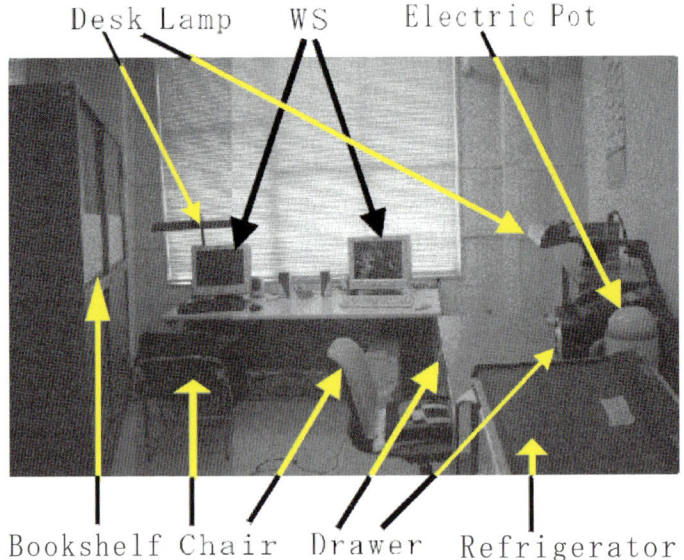

Fig. 4.1 Vivid Room.

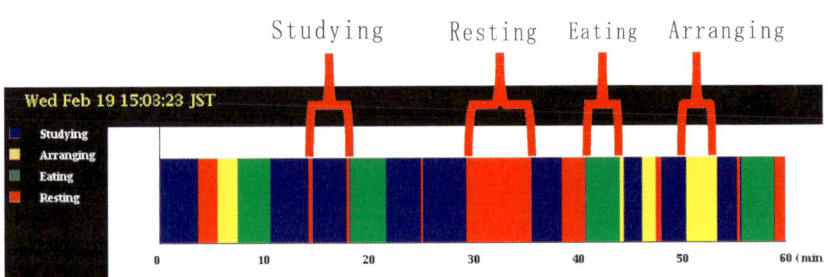

Fig. 4.2 An example of recognized intentions in Vivid Room.

In all the above-mentioned research, certain types of human activities are recognized. However, even if we know that a human is cooking something, systems supporting human activity are rather limited. Suppose that a human is making a cup of coffee, it will be useful to suggest where the cream is, when the human picked up a cup and the coffee. In order to provide such suggestions, the system should know the time series of procedures and infer the next action that would be executed by the human based on past and present observed actions.

In this chapter, we introduce a human activity recognition system, which infers the next human action by taking account of past human behaviors. All the merchandise in supermarkets has a one-dimensional bar code. But in near the future, they will be replaced by IC tags, which maintain information such as manufacturer name, type of merchandise, place of production, expiration date, etc. This means that all items in a home will be labeled with IC tags. So in this paper, we presume foods, cooking tools, tableware, and cutlery in the kitchen are labeled with IC tags and the movements of these items can be tracked with the use of antennas placed on shelves and kitchen counters.

We developed human a activity support system for a kitchen using an LCD touch panel on the kitchen counter and a mobile robot on the floor. The panel displays the recipes with pictures. The robot suggests the next action human should take using voice and gestures. We also conducted experiments to verify the adequacy and the quality of activity support.

4.2 System Design

4.2.1 Inference from Series of Human Actions

At first, we discuss the characteristics of human behaviors we need to recognize. We define an observed action by sensors as **action** a_i and a set of actions as $A = \{a_1, a_2, ..., a_n\}$. For example, a set of action sequence will be a_1: "take a cup from the cupboard", a_2: "take an instant coffee from the cupboard", a_3: "take a spoon from the drawer", and a_4: "take an electric pot for pouring hot water" when someone is making a cup of coffee.

We also define a set of time series actions in arbitrary length as **action pattern** p_i and a set of action patterns as $P = \{p_1, p_2, ..., p_m\}$. We have observed the action pattern $p_o = \{\mathbf{a_3}, \mathbf{a_2}, \mathbf{a_4}\}$ by watching a human and there exists an action pattern $p_i = \{..., \mathbf{a_3}, \mathbf{a_2}, \mathbf{a_4}, a_6, ...\}$ in database P, which is the collection of action patterns observed so far.

We could find the same time series action pattern in p_i and can infer that the next action the human should execute is action a_6.

Humans sometimes behave redundantly or concurrently. For example, p_o may contain a_n as $\{\mathbf{a_3}, \mathbf{a_2}, a_n, \mathbf{a_4}\}$ or p_i may contain a_n as $\{..., \mathbf{a_3}, a_n, \mathbf{a_2}, \mathbf{a_4}, a_6, ...\}$. These actions are considered as noise when the original time series actions consist of the actions for making coffee, cooking hamburgers, etc. So we must develop the inference system, which could infer the next human action even if such noises are contained.

Add to that, human procedures (time series actions) may have branches. For example, one may add sugar and the other may add cream after he/she has made black coffee. These phenomena mean that there may exist $p_j = \{..., a_3, a_2, a_4, a_7, ...\}$ in addition to p_i mentioned above. If the inference system uses not only time series information but also frequency of action patterns observed so far, it can predict the preferred next action the human should perform. For example, when you see your familiar person who always drinks black coffee has added cream in it, you will easily be aware of it.

4.2.2 Time Sequence Data Mining

The data mining methods can be roughly categorized into four kinds (i.e., correlation analysis, time sequence analysis, clustering, and learning) [6]. The time sequence analysis is the one which is suited to our purpose. In this section, we explain briefly the typical algorithms that extract temporal orders in time series patterns.

The Apriori algorithm proposed by Agrawal is one of the famous data mining methods for temporal sequential data [2]. We will explain the a priori algorithm by using examples. Supposed that there are four time series data sets $p_1 = \{a_3, a_2, a_4, a_6\}, p_2 = \{a_3, a_2, a_4, a_6\}, p_3 = \{a_3, a_2, a_5, a_6\}, p_4 = \{a_3, a_1, a_4, a_6\}$ in a database. The Apriori algorithm extracts the partial sequences from the data sets by taking into account the number of occurrences and certainty given by a user. For example, it finds the partial sequences $\{a_3, a_4\}$, which means "a_4 occurs after a_3". The certainty is the occurrence ratio (i.e., a_4 happens after a_3 at the ratio of 75%). It is known experimentally that the calculation costs increase exponentially as the number of data sets increases with the a priori algorithm.

On the other hand, Pei proposed the PrefixSpan algorithm, which extracts multiple frequency patterns efficiently in terms of computational costs [14]. Suppose that there are four time series data sets $p_1 = \{a_3, a_2, a_4, a_6\}, p_2 = \{a_3, a_2, a_4, a_6\}, p_3 = \{a_3, a_2, a_5, a_6\}, p_4 = \{a_3, a_1, a_4, a_6\}$ in the same as the above example. PrefixSpan extracts the partial sequences with the number of occurrences as shown in Figure 4.3. $\{a_3/4, a_2/3, a_4/2, a_6/2\}$ in the figure denotes that there is a time series data $\{a_3, a_2, a_4, a_6\}$ with the frequency shown as suffix (i.e., the occurrence of a_3 alone is 4 and the occurrence of $\{a_3, a_2, a_4, a_6\}$ as the time series data is 2).

4.2.3 Human Behavior Inference Algorithm

Since the above-mentioned time sequence data mining methods allow exact matching, we will propose a human behavior inference algorithm that takes into

$a_1/1$	$a_4/1$	$a_6/1$	
$a_1/1$	$a_6/1$		
$a_2/3$	$a_4/2$	$a_6/2$	
$a_2/3$	$a_5/1$	$a_6/1$	
$a_2/3$	$a_6/3$		
$a_3/4$	$a_1/1$	$a_4/1$	$a_6/1$
$a_3/4$	$a_1/1$	$a_6/1$	
$a_3/4$	$a_2/3$	$a_4/2$	$a_6/2$
$a_3/4$	$a_2/3$	$a_5/1$	$a_6/1$
$a_3/4$	$a_2/3$	$a_6/3$	
$a_3/4$	$a_4/3$	$a_6/3$	
$a_3/4$	$a_5/1$	$a_6/1$	
$a_3/4$	$a_6/4$		
$a_4/3$	$a_6/3$		
$a_5/1$	$a_6/1$		
$a_6/4$			

Fig. 4.3 Time series data generated by PrefixSpan.

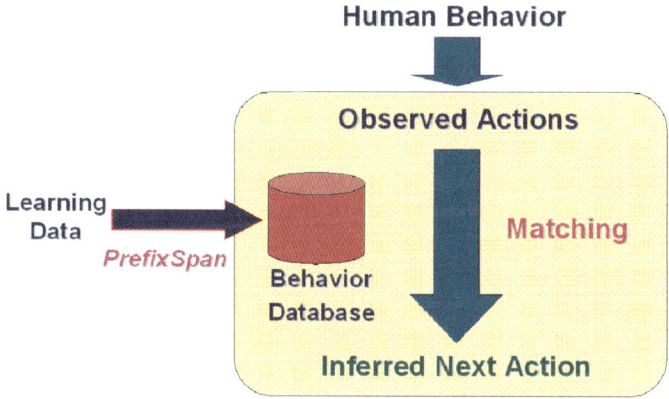

Fig. 4.4 Overview of proposed inference engine.

consideration noise both in the time sequence data within the database and in the human observation data .

The method to predict the human's next behavior is to find a time sequence data that is the same as the window data from the behavior database. The database contains a large amount of time sequence data generated from observations of past behaviors. We employ PrefixSpan for generating partial time sequence data from the behavior database because PrefixSpan has lower computational costs and easy to utilize compared with other similar tools.

A diagram of the proposed inference engine is shown in Figure 4.4. We developed the window-based matching method that is insensitive to noise. We also employ a certainty measure to indicate the confidence of the inferred results.

First, we will define the terms used in the algorithm. Every behavior observed by sensors is defined as **input data** w_i. Suppose that the latest input data is w_i and the number W of input data recently observed is $\{w_{i-W+1}, \cdots, w_i\}$; these W time series data are defined as **window data** of width W.

The matching between the input data and the behavior database is done by window size. For example, if we start to find the time series input data of window size 5 and could not find an exact match, we will reduce the window width to $4, 3, 2$. In this way, even if the input data contains some noise, we will be able to find an exact match in the database. The maximum window size used at the beginning of the search is defined as W_{max} and the minimum window size is defined as W_{min}. In order to infer the human's next action, several time sequence data are required and so W_{min} is used for terminating the search.

The **inferred event** is the action that succeeds the matched time sequence with window size W at the highest certainty. The **certainty** is calculated as follows:

$$certainty = \frac{O_{ia}}{O_{pa}}, \quad (0 \leq certainty \leq 1) \tag{4.1}$$

where O_{ia} is the Occurrence of the Inferred Action and O_{pa} is the Occurrence of the Preceding Action to the Inferred Action calculated by PrefixSpan. The occurrence is the number of events observed in a learning instance. In the example shown in Figure 4.5, since both O_{ia} and O_{pa} are 2, the certainty is calculated as $100\% (= \frac{2}{2})$.

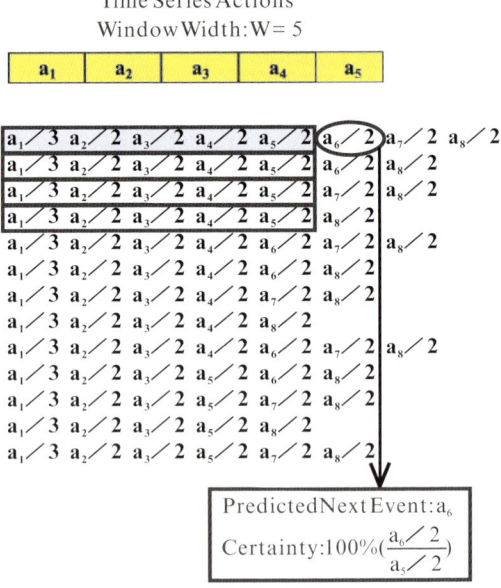

Fig. 4.5 Matching algorithm.

The outputs of the inference engine are one of the following three kinds: EOS (End Of Sequence), inferred event with certainty, or NULL. EOS is the output when the observed time series data was matched in the database but the observed most recent event is the end of sequence in the matched database. So the inference engine could not infer the next event in this case. When the inference engine found the matched time series data and there was a succeeding event in the database, it outputs the succeeding event as the inferred event with the certainty calculated by formula 4.1. NULL is the output when the inference engine could not find the matched data in the database even though it reduced the time series actions window width to W_{min}.

The overall algorithm of our proposed inference engine is as follows:

1) At first, it creates the window with the size $W = 5$ and initializes all contents as NULL.

2) Then, it inputs the observed events at most $W = 5$ to the window so that the most recent event becomes w_W.

3) It finds the time series data that is the exact match of the window in the database.

4) If there was only one match, it outputs the inferred event ($(W + 1)$th event in the database) with its certainty calculated by Equation 4.1. If $(W + 1)$th event does not exist, it outputs EOS.

5) If there was more than one match, it selects all matches with a high certainty, then chooses the longest sequences[1]. If there remain multiple candidates after applying the strategy mentioned above, it selects a candidate arbitrarily and outputs the inferred event or EOS as in step 4.

6) If there were no matched data from step 3, it reduces the window size to $W = W - 1$ as shown in Figure 4.6. It then finds the matched data with these multiple windows as step 3. When the window size becomes $W < W_{min}$, it outputs NULL since it could not find a match in the database. The above-mentioned procedure is summarized in Table 4.1.

4.2.4 Activity Support of Human

If the system suggests the next action to be performed, it would be very helpful. At the same time, it is important not to perturb a human's free activities based on his/her own preferences. So in this research, we develop a mobile robot that recommends the human's next action by voice and gesture.

In order to minimize unhelpful or unsuitable recommends, we employed a threshold for certainty (see formula 4.1). The robot would only suggest a next action when the certainty of inferred event is above the threshold.

---

[1] This is because the longer the time series data, the better it describes the detailed procedures.

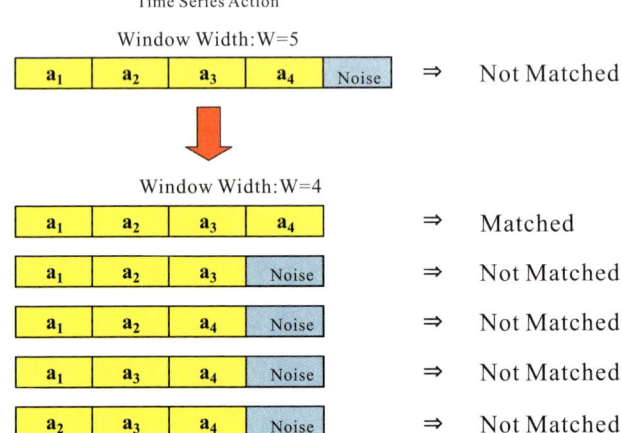

Fig. 4.6 Reduction of window size.

Table 4.1 The procedure of matching algorithm.

No.	Procedures
1	Set $W = 5$ and creates the window $\{w_1, w_2, w_3, w_4, w_5\}$ with all values as NULL.
2	Inputs the observed at most $W = 5$ events to the window so that the most recent event becomes w_W.
3	Finds the exact same time series events as in the window in the database. No. of matched data is singular. $\Rightarrow$ go to 4. No. of matched data is plural. $\Rightarrow$ go to 5. No. of matched data is none. $\Rightarrow$ go to 6.
4	Outputs the inferred event ($(W+1)$th event in the database) with its certainty calculated by formula 4.1. Outputs EOS if $(W+1)$th event does not exist.
5	Selects the highest certainty ones, then selects the longest sequence of ones.
6	Reduces the window size to $W = W - 1$. $\Rightarrow$ go to 3. If the window size becomes $W < W_{min}$, it outputs NULL.

   We conducted the experiments with 10 subjects and collected the subjects' comments regarding the usefulness of the recommendations (inferred events). The results suggest that the certainty below which the subjects felt that the recommendations were unsuitable was 0.55. So we have set the threshold to 0.55; the robot will not issue recommendations whose certainty is below this value.

4.3 Implementation

4.3.1 IC Tag System

We presume that all the merchandise in supermarkets and department stores will have IC tags in the near future, replacing bar codes. Thus in the future most of the items in a house and office will have IC tags, enabling location and movement tracking using antennae.

In this work, we employed the IC tag system developed by Feig Electronics Co. Ltd. The size of the IC tag (sticker label) is approximately 2 cm × 5.5 cm. The size of the antenna is approximately 30 cm × 40 cm and it can read/write the information from/to the IC tags at a distance up to approximately 15 cm.

We have attached IC tags to the items (cup, glass, pot, instant coffee, tea bag, cream, sugar, potato, carrot, spoon, fork, knife, medicine box, disinfectant, cotton, adhesive plaster, etc.), which are usually available in a kitchen or home (see Figure 4.7).

Information from the tag is downloaded to the PC via an RS-232C serial link. The inference engine is implemented on the PC and the inferred event is transferred to the mobile robot via wireless LAN (see Figure 4.8).

Fig. 4.7 Items labeled by IC tags.

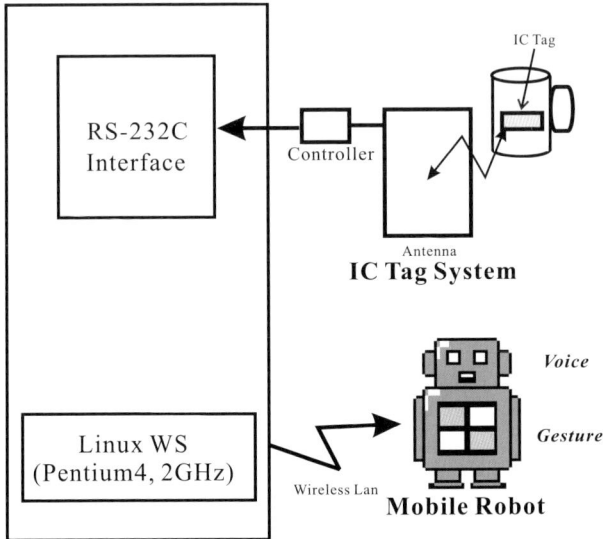

Fig. 4.8 System configuration of cooking support robot.

```
Cup-a0 Pot-a0 Pot-a1 TeaBag-a0 TeaBag-a1 Sugar-a0 Sugar-a1 Cream-a0 Cream-a1 Spoon-a0
Cup-a0 Pot-a0 Pot-a1 TeaBag-a0 TeaBag-a1
Cup-a0 Pot-a0 Pot-a1 TeaBag-a0 TeaBag-a1
Cup-a0 Pot-a0 Pot-a1 TeaBag-a0 TeaBag-a1
Cup-a0 Pot-a0 Pot-a1 TeaBag-a0 TeaBag-a1
Cup-a0 Pot-a0 Pot-a1 TeaBag-a0 TeaBag-a1
Cup-a0 Pot-a0 Pot-a1 TeaBag-a0 TeaBag-a1 Lemon-a0 Lemon-a1 Spoon-a0
Cup-a0 TeaBag-a0 Pot-a0 Pot-a1 TeaBag-a1 Sugar-a0 Sugar-a1 Spoon-a0
Cup-a0 TeaBag-a0 Pot-a0 Pot-a1 TeaBag-a1
Cup-a0 TeaBag-a0 Pot-a0 Pot-a1 TeaBag-a1 Lemon-a0 Lemon-a1
```

Fig. 4.9 Example of learning data.

4.3.2 Inference of Human's Next Action

In order to obtain the learning instances for PrefixSpan, we asked 10 subjects to perform five kinds of tasks, which are 1) make a cup of coffee, 2) make a cup of tea, 3) treat a cut on a finger, 4) take a medicine for a cold, and 5) make a curry and rice. The examples of the learning instances are as shown in Figure 4.9.

In order to predict precise human behaviors, we have employed not only the IC tag information that provides the name of the item, but also location information i.e., where the item was sensed (a: cupboard, b: cabinet, c: medicine box) and the human action (0: taken out, 1: stored). For example, the event Spoon-a0 denotes that the spoon was taken from the cupboard.

```
Cup-a0/20 Pot-a1/10 TeaBag-a1/10 Cream-a0/1 Cream-a1/1 Spoon-a0/1
Cup-a0/20 Pot-a1/10 TeaBag-a1/10 Cream-a0/1 Spoon-a0/1
Cup-a0/20 Pot-a1/10 TeaBag-a1/10 Cream-a1/1 Spoon-a0/1
Cup-a0/20 Pot-a1/10 TeaBag-a1/10 Lemon-a0/2 Lemon-a1/2 Spoon-a0/1
Cup-a0/20 Pot-a1/10 TeaBag-a1/10 Lemon-a0/2 Spoon-a0/1
Cup-a0/20 Pot-a1/10 TeaBag-a1/10 Lemon-a1/2 Spoon-a0/1
Cup-a0/20 Pot-a1/10 TeaBag-a1/10 Spoon-a0/3
Cup-a0/20 Pot-a1/10 TeaBag-a1/10 Sugar-a0/2 Cream-a0/1 Cream-a1/1 Spoon-a0/1
Cup-a0/20 Pot-a1/10 TeaBag-a1/10 Sugar-a0/2 Cream-a0/1 Spoon-a0/1
Cup-a0/20 Pot-a1/10 TeaBag-a1/10 Sugar-a0/2 Cream-a1/1 Spoon-a0/1
Cup-a0/20 Pot-a1/10 TeaBag-a1/10 Sugar-a0/2 Spoon-a0/2
Cup-a0/20 Pot-a1/10 TeaBag-a1/10 Sugar-a0/2 Sugar-a1/2 Cream-a0/1 Cream-a1/1 Spoon-a0/1
Cup-a0/20 Pot-a1/10 TeaBag-a1/10 Sugar-a0/2 Sugar-a1/2 Cream-a0/1 Spoon-a0/1
Cup a0/20 Pot-a1/10 TeaBag-a1/10 Sugar-a0/2 Sugar-a1/2 Cream-a1/1 Spoon-a0/1
Cup-a0/20 Pot-a1/10 TeaBag-a1/10 Sugar-a0/2 Sugar-a1/2 Spoon-a0/2
Cup-a0/20 Pot-a1/10 TeaBag-a1/10 Sugar-a1/2 Cream-a0/1 Cream-a1/1 Spoon-a0/1
Cup-a0/20 Pot-a1/10 TeaBag-a1/10 Sugar-a1/2 Cream-a0/1 Spoon-a0/1
Cup-a0/20 Pot-a1/10 TeaBag-a1/10 Sugar-a1/2 Cream-a1/1 Spoon-a0/1
Cup-a0/20 Pot-a1/10 TeaBag-a1/10 Sugar-a1/2 Spoon-a0/2
```

Fig. 4.10 Example of time series data generated by PrefixSpan.

In the current implementation, we are using only one antenna thus forcing users to scan the items on the antenna. The system knows that the tagged object was taken when the IC tag is sensed by the antenna the first time and recognizes that the object is stored when the same IC tag is sensed a second time. The storage locations of the items are predefined and they are hard-coded. But in the future, by installing IC tag antennas on each of the shelves and kitchen counters, the system will locate the items in realtime without the need for users to scan.

The time sequence database generated by PrefixSpan from the learning data shown in Figure 4.9 is as shown in Figure 4.10. For example, the data {Cup-a0/20, Pot-a1/10, TeaBag-a1/10, Spoon-a0/3} denotes that the event Cup-a0 on its own was observed 20 times in the learning data. But the sequence of {Cup-a0, Pot-a1, TeaBag-a1, Spoon-a0} (the cup was taken out of the cupboard, the pot was stored in the cupboard, the tea bag was stored in the cupboard, and the spoon was taken out of the cupboard) was observed 3 times.

4.3.3 Cooking Support Interface

We employed the Robovie mobile robot developed at ATR [7] as the cooking-support robot (see Figure 4.11). We programmed the robot to recommend the inferred next human action using synthesized voice and gestures.

We also installed an LCD touch panel on the wall of kitchen counter. It displays recipes with instructions as shown in Figure 4.12. In the left frame, step number of the cooking procedures is shown. In the right frames, two steps of detailed

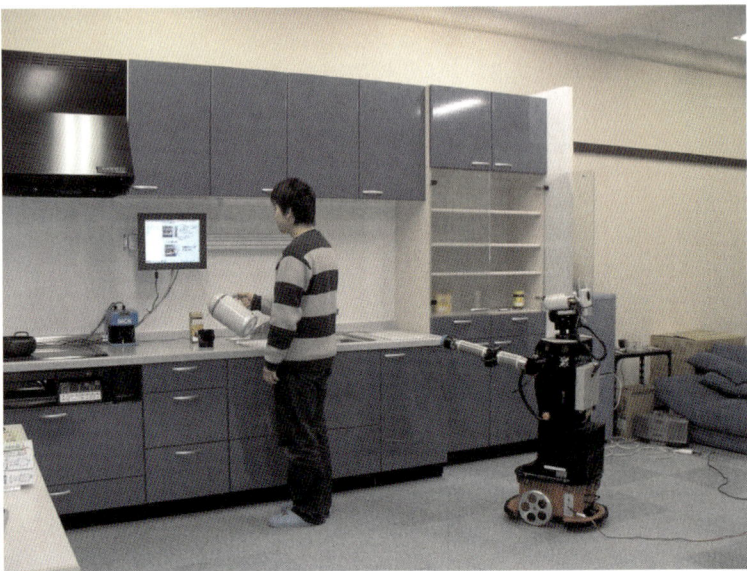

Fig. 4.11 Cooking-support robot recommending a presumed next action to human by voice and gesture.

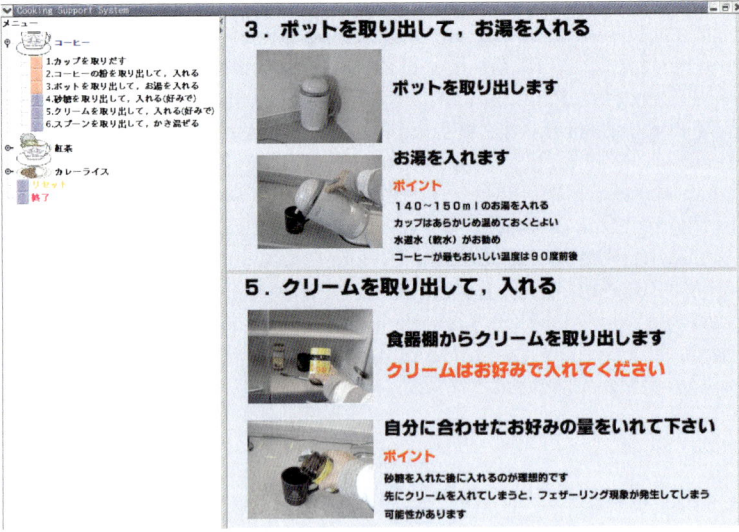

Fig. 4.12 Recipe shown on LCD.

instruction with pictures is shown. For example, when the system recognized a pot picked by using the IC tag, the instruction for pouring hot water to make coffee will be shown on the right-upper frame and the instruction of the succeeding step on the right-bottom frame. When the system infers the next action, the panel displays the instruction for the inferred one on the right-bottom frame. The example shown in Figure 4.12 is the case where the system inferred the next step as putting cream in the cup. Automatic scrolling allows users to browse through steps of the recipe by touching the step numbers displayed in the left frame.

We confirmed that the following supports have been realized. When a user took out a cup and instant coffee from the cupboard, the robot recommends the next action by saying "sugar is in the cupboard" and by turning toward the cupboard and pointing the shelf where the sugar is located. When a user took cold medicine and stored it in the medicine box, the robot recommends the next action by saying "the medicine box should be stored on the shelf" and pointing to the shelf. These recommendations are automatically generated from the inferred events such as Sugar-a0, MedicineBox-b1, etc.

4.4 Experimental Results

To evaluate the adequacy and the quality of action support, we conducted experiments with 10 subjects other than those used for collecting learning instances. We instructed new subjects to speak out the short phrases according to the adequacy they feel, each time they hear suggestions from the robot. The phrases we instructed were as shown in Table 4.2 and they are scored from -1 to 1. We videotaped the experiments and counted the scores each time the robot made a suggestion.

First to confirm the adequacy of suggestions in each task (tasks 1 to 4 explained in Section 4.3.2), we asked subjects to perform each of the tasks. The averaged scores are shown in Figure 4.13 and all scores were greater than 0.8. From the results, we confirmed that the system could infer a suitable next action and the suggestion made by the robot were accepted by the subjects. There were no incorrect suggestions observed during the experiments. Some suggestions such as recommending the use of a spoon for stirring were ignored (scored as 0), because some users did not use a spoon, e.g., for black tea.

In order to evaluate the robustness to noise in the time series of observed, we instructed a pair of subjects to perform different tasks. For example, we instructed one subject to make a cup of coffee and the other to take a medicine for a cold. Their actions (usages of items) will be interleaved and the actions of one subject are noise to the other. The averaged scores were as shown in Figure 4.14.

Table 4.2 Phrases used for evaluation.

Score	Phrase
1	"Thank you."
0	(silent/ignore)
-1	"No thanks."

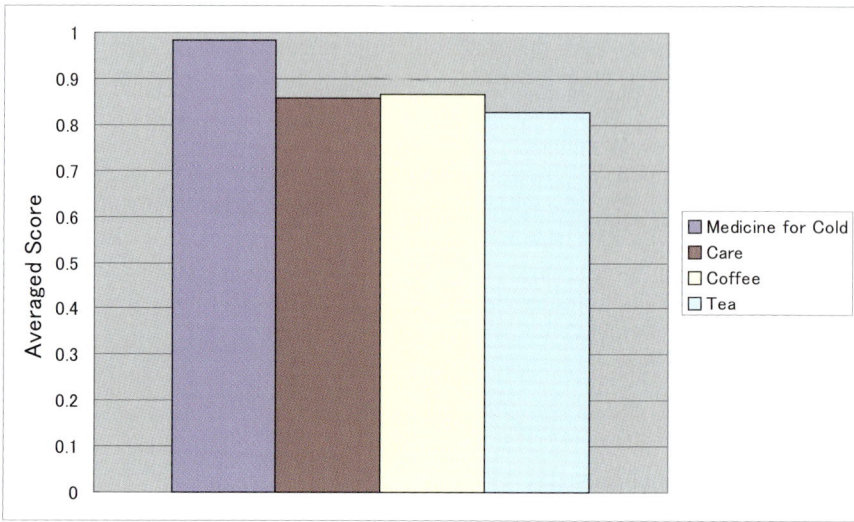

Fig. 4.13 Evaluation of adequacy in each task.

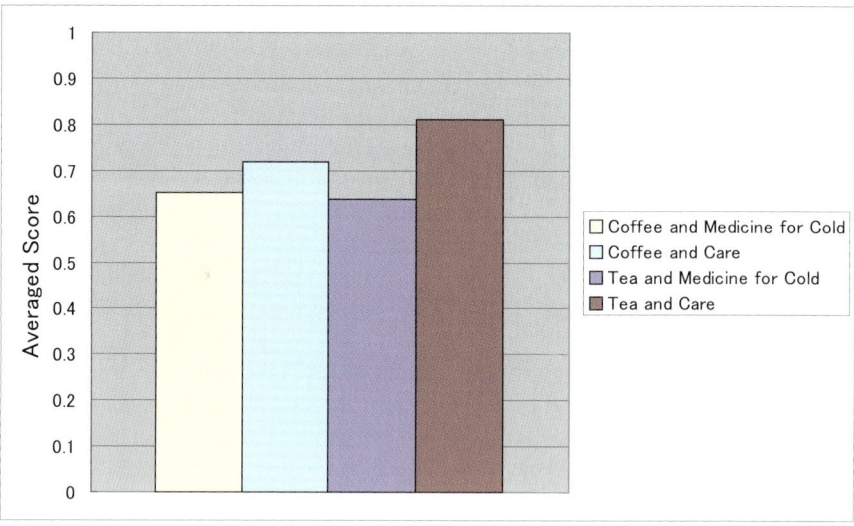

Fig. 4.14 Evaluation of adequacy when two tasks were interleaved.

Even though the two tasks were interleaved, the averaged scores were still high (about 0.7). From the results, we confirmed that the reduction of window size (explained in section 4.2.3) was effective to make the system robust to noise in observed action sequences. Again, there were no incorrect suggestions observed during the experiments. Since some recommendations by the system are for the other subject,

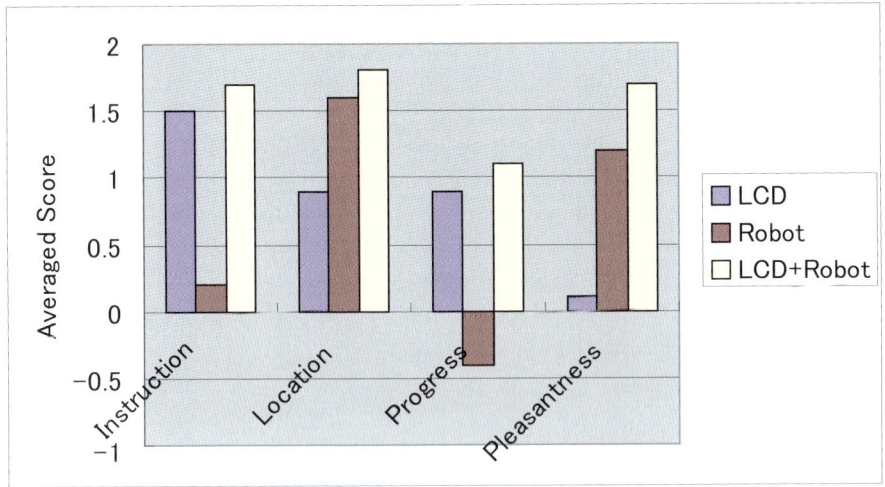

Fig. 4.15 Evaluation of the support by LCD and mobile robot.

those are just ignored (scored as 0). The subjects understood that those recommendations were for the subject working nearby.

In order to evaluate the adequacy of two types of supporting media, we asked subjects to make a cup of coffee under three conditions (i.e., (1) support by LCD, (2) support by robot, and (3) support by both LCD and robot). After the task, we asked the subject to score (from −2 to 2) the quality of support in terms of (a) comprehensibility of instructions, (b) comprehensibility of object locations, (c) comprehensibility of progress within the recipe, and (d) pleasantness of task.

As shown in Figure 4.15, it was confirmed that the instruction supported with the LCD is suitable for understanding instructions and progress. On the other hand, the instructions given by the robot are suitable for indicating the location of objects and increase the pleasantness of the work. In all cases, the combined use of two media increases the quality of human support over that of a single medium.

4.5 Conclusions

In this paper, we propose a human behavior recognition system which infers the typical next human action by taking into account the accumulated human behaviors observed in the past. We also developed a cooking support system by using an LCD touch panel and a mobile robot, which recommends the presumed next human action. From experimental results with subjects, we confirmed the adequacy of the proposed inference system and the quality of support.

The characteristics of the proposed recognition system are as follows. (1) It is robust to noise both in the time sequence data within the database and in the human observation data. These noises are inevitable for the system that allows user's free activities in intelligent spaces. The robustness is confirmed as the experimental results. (2) The certainty is calculated with the inferred next action. (3) The data mining method is employed. Therefore, the system could be adapted to different types of applications by expanding the data. If we import recipes conducted by professional cooks, it will be attractive for both novice cooks as well as skilled cooks.

In future work, the system will suggest a recipe taking into account the foods available in the kitchen. We also plan to extend the system so that it detects more precise and detailed human activities by using heterogeneous sensors such as vision, laser, etc.

References

1. K. Asaki, Y. Kishimoto, T. Sato, and T. Mori, "One-Room-Type Sensing System for Recognition and Accumulation of Human Behavior –Proposal of Behavior Recognition Techniques," *Proc. of JSME ROBOMEC'00*, 2P1-76-119, 2000.
2. R. Agrawal and R. Strikant, "Fast algorithms for mining association rules," *Proc. of the 20th International Conference on Very Large Databases*, pp.487–499, 1994.
3. B. Brumitt et al., "Easy Living: Technologies for Intelligent Environments," *Proc. of International Symposium on Handheld and Ubiquitous Computing*, 2000.
4. B. Brumitt, B. Meyers, J. Krumm, A. Kern, and S. Shafer, "EasyLiving: Technologies for Intelligent Environments," *Proc. of International Symposium on Handheld and Ubiquitous Computing*, pp.12–29, 2000.
5. I.A. Essa, "Ubiquitous sensing for smart and aware environments: technologies towards the building on an aware home," *Position Paper for the DARPA/NFS/NIST workshop on Smart Environment*, 1999.
6. U.M. Fayyad, G. Piatetsky-Shapiro, P. Smyth, and R. Uthurusamy, "Advances in Knowledge Discovery and Data Mining," MIT Press, 1996.
7. H. Ishiguro, T. Ono, M. Imai, T. Maeda, T. Kanda, and R. Nakatsu, "Robovie: A robot generates episode chains in our daily life," *Proc. of Int. Symposium on Robotics*, pp.1356–1361, 2001.
8. J. Krumm, S. Harris, B. Meyers, B. Brumitt, M. Hale and S. Shafer, "Multi-Camera Multi-Person Tracking for Easy Living," *Proc. of 3rd IEEE International Workshop on Visual Surveillance*, pp.3–10, 2000.
9. J. Lee, N. Ando, and H. Hashimoto, "Design Policy for Intelligent Space," *Proc. of IEEE International Conference on System, Man and Cybernetics (SMC'99)*, pp.12–15, 1999.
10. D.J. Moore, I.A. Essa, and M.H. Hayes III, "ObjectSpaces: Context Management for Human Activity Recognition," Georgia Institute of Technology, Graphics, Visualization and Usability Center, Technical Report, #GIT-GVU-98-26, 1998.
11. D.J. Moore, I.A. Essa, and M.H. Hayes III, "Exploiting Human Actions and Object Context for Recognition Tasks," *Proc. of the 7th IEEE International Conference on Computer Vision*, pp.80–86, 1999.
12. T. Mori, T. Sato et al., "One-Room-Type Sensing System for Recognition and Accumulation of Human Behavior," *Proc. of IROS2000*, pp.345–350, 2000.
13. Y. Nakauchi et al., "Vivid Room: Human Intention Detection and Activity Support Environment for Ubiquitous Autonomy," *Proc. of IROS2003*, pp.773–778, 2003.
14. J. Pei et al., "PrefixSpan: Mining Sequential Patterns Efficiently by Prefix-Projected Pattern Growth," *Proc. of International Conference of Data Engineering*, pp.215–224, 2001.
15. A. Pentland, "Smart Rooms," *Scientific American*, pp.54–62, 1996.

16. A. Pentland, R. Picard, and P. Maes "Smart Rooms, Desks, and Clothes: Toward Seamlessly Networked Living," *British Telecommunications Engineering*, Vol.15, pp.168–172, July, 1996.
17. T. Sato, Y. Nishida, and H. Mizoguchi, "Robotic Room: Symbiosis with human through behavior media," *Robotics and Autonomous Systems 18 International Workshop on Biorobotics: Human-Robot Symbiosis*, Elsevier, pp.185–194, 1996.
18. M. Weiser, "The Computing for the Twenty-First Century," *Scientific American*, pp.94–104, September, 1991.

Chapter 5
Robotic, Sensory and Problem-Solving Ingredients for the Future Home

Amedeo Cesta[a], Luca Iocchi[b], G. Riccardo Leone[a,b], Daniele Nardi[b], Federico Pecora[a], and Riccardo Rasconi[a]

Abstract ROBOCARE has been a three-year Italian research project aimed at assessing the extent to which different state-of-the-art technologies can benefit the creation of an assistive environment for elder care. In its final year, the project focused on producing a demonstration exhibiting an integration of robotic, sensory and problem-solving software agents. This article describes the ROBOCARE Domestic Environment, an experimental three-room flat in which a number of heterogeneous robotic, domotic, and intelligent software agents provide domestic cognitive support services for elderly people. The RDE is a deployed multiagent system in which agents coordinate their behavior to create user services such as nonintrusive monitoring of daily activities and activity management assistance. This article provides a summary of the system's key features, focusing on the integrated prototypical environment which was deployed in the ROBOCARE lab in Rome and exhibited at the RoboCup 2006 competition.

5.1 Introduction

ROBOCARE has been a three-year research project[1] which aimed at developing multiagent systems for the care of the aging population. The principal aim of ROBOCARE is to assess the extent to which different state-of-the-art technologies can benefit the creation of an assistive environment for elder care, one of its main driving forces being the increasing abundance of "intelligent" domestic devices and

[a] Institute for Cognitive Science and Technology, National Research Council of Italy, `<amedeo.cesta>@istc.cnr.it` ·
[b] Dipartimento di Informatica e Sistemistica, University of Rome "La Sapienza", Italy, `nardi@dis.uniroma1.it`

[1] This research was sponsored by MIUR (Italian Ministry of Education, University and Research) under project ROBOCARE (A Multi-Agent System with Intelligent Fixed and Mobile Robotic Components), L. 449/97.

affordable pervasive computing technology. It is with this aim that the final year
of the project focused on producing a demonstration exhibiting an integration of
robotic, sensory and problem-solving software agents. To this end, an experimental setup which re-creates a three-room flat was set up at the ISTC-CNR in Rome,
named The ROBOCARE Domestic Environment (RDE). The RDE is intended as a
testbed environment in which to test the ability of heterogeneous robotic, domotic,
and intelligent software agents to provide cognitive support services for elderly people at home. Specifically, the RDE is a deployed multiagent system in which agents
coordinate their behavior to create user services such as nonintrusive monitoring of
daily activities and activity management assistance. A key feature of the RDE is a
context-aware domestic robot developed by the RoboCare team at the ISTC-CNR[2].

The robot is aimed at demonstrating the feasibility of an embodied interface between the assisted elder and the smart home. Thus, the RDE can be viewed as a "ro-
botically rich" environment composed of sensors and software agents whose overall
purpose is to (a) predict/prevent possibly hazardous behavior; (b) monitor the adherence to behavioral constraints defined by a caregiver; (c) provide basic services
for user interaction.

The system was partially re-created in the RoboCup@Home domestic environment during the RoboCup 2006 competition in Bremen[3] where it was awarded third
prize.

5.1.1 Components of the Multiagent System

The RDE is equipped with the following agents, which provide services of various
nature:

- A domestic service robot, endowed with laser-based scan matching algorithms
  for robust self-localization, and with path planning and obstacle avoidance algorithms (see Figure 5.1).
- An Interaction Manager which coordinates Speech-act synthesis and user feedback interpretation, therefore providing an intuitive User Interface to/from the
  robot.
- Two fixed stereo cameras providing a People Localization and Tracking (PLT)
  service, and a Posture Recognition (PR) service.
- An ADL (Activities of Daily Living) monitor, a scheduling and execution monitoring system which is responsible for monitoring the assisted person's daily
  activities and assessing the adherence to behavioral constraints defined by a caregiver.

---

[2] The development team is the result of a combined development effort stemming from two partners
of the RoboCare project, namely, the Planning and Scheduling Team at ISTC-CNR and SPQR at
the University of Rome "La Sapienza".

[3] Competition homepage: http://www.ai.rug.nl/robocupathome/.

- One personal data assistant (PDA) on which a very simple four-button interface is deployed. The interface allows to (1) summon the robot, (2) send the robot to a specific location, (3) relay streaming video from the robot to the PDA, and (4) stop the robot.

The robotic mediator was built to explore the added value of an embodied companion in an intelligent home. Its mobility also provides the basis for developing a number of added-value services which require physical presence. Because of the tasks that the robot has to accomplish in the environment, its localization and mapping capabilities have great importance and will be described in Section 5.2.

The Interaction Manager (IM) is the module that coordinates the services provided by a Voice Recognition agent and by a Speech Synthesis agent; thanks to the IM, simple natural language bidirectional communication between the robot and the user can be established. The IM will be briefly discussed in Section 5.3.

The Stereo-vision-based People Localization and Tracking service (PLT) provides the means to locate the assisted person. The system is scalable as multiple cameras and can be used to improve area coverage and precision: in addition, vision-based Posture Recognition (PR) can be cascaded to the PLT computation in order to provide further information on what the assisted person is doing. The sensory subsystem is described in Section 5.4.

Continuous feedback from the sensors allows to build a symbolic representation of the state of the environment and of the assisted elder. This information is employed by a CSP-based schedule execution monitoring tool (T-REX [2, 1]) to follow the occurrence of Activities of Daily Living (ADLs). The aspects of daily life to be monitored are specified by a caregiver in the form of complex temporal constraints among activities. Constraint violations lead to system intervention (e.g., the robot suggests "how about having lunch?", or warns "don't take your medication on an empty stomach!"). The details of the ADL monitor are shown in Section 5.5.

Overall, the RDE is a collection of service-providing components of various nature. Sensors contribute to building a symbolic representation of the state of the environment and of the assisted person. Based on this information, automated reasoning agents infer actions to be performed in the environment, principally through the robotic mediator. Both enactment and sensing require the synergistic cooperation of multiple capabilities from different agents, such as robot mobility, speech synthesis and recognition, and so on. For this reason, multiagent coordination is an important aspect of the RDE scenario. Section 5.6 is dedicated to the description of the coordination mechanism, which occurs in the RDE's current configuration by means of ADOPT-N [3], a distributed constraint reasoning algorithm.

5.2 The Robotic Platform Mobility Subsystem

This present Section briefly describes the functionalities of navigation, path planning, mapping and localization providing the basis for added-value services which require physical presence.

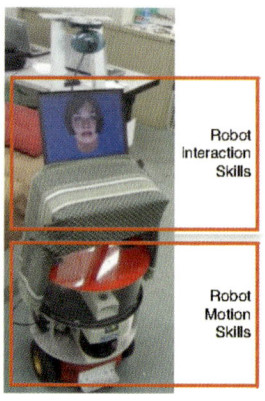

Fig. 5.1 The Robot's Mobility and Interaction modules.

Localization and mapping is the primary requirement for implementing a robust mobile platform in the domestic environment. Underlying the mobility services is a Sampling Importance Resampling (SIR) particle filtering algorithm, which is extensively described in [4]. A significant part of our research in the early stages of the ROBOCARE project was dedicated to obtaining a reliable and robust mobility subsystem for the robotic mediator. The results of this research are a set of key mobility services consisting in primitives which can be invoked to make the robot reach any position in the domestic environment. In particular, SIR is particularly suited for the domestic scenario, in which the map of the environment may change in an unpredictable manner. Indeed, the approach allows to take into account the position of chairs, tables, sofas, or any other object whose position is likely to change over time.

Given the capability of localizing itself in the environment, the mobile platform must provide a "goto-place" service which can be invoked in order to make the robot move robustly from one position in the environment to the other. In particular, the ROBOCARE robotic platform provides two levels of mobility services: a `goto-XY(x,y)` function on the one hand, which triggers the robot to reach a certain (x,y) position in the environment, and a `goto-place(dest)` primitive through which the robot can be sent to a particular known destination (such as "the sofa", or "the lamp"). Clearly, the latter functionality is at a higher level of abstraction than the former, and in our system consists in a naming scheme which associates names to coordinate pairs. Therefore, invoking the `goto-place()` command will result in a look-up in the location database followed by the appropriate invocation of the `goto-XY()` functionality. Since the core of the mobility infrastructure comes into play at the `goto-XY()` invocation level, we here briefly describe the topological path planning algorithm underlying this primitive.

Autonomously navigating toward a given coordinate pair in the domestic setting is not a trivial problem. It poses both general problems pertaining to autonomous navigation, as well as problems which are unique to the domestic environment. Using complete algorithms to find the topology of the environment (e.g., Voronoi diagrams) is very expensive and, since we have a different map at each cycle, a probabilistic approach is more convenient for the topological path-planner.

The most widely used approach that builds a graph representing a roadmap of the environment is the Probabilistic RoadMap (PRM) [5] algorithm. This algorithm works by picking random positions in the configuration space and trying to connect them with a fast local planner. The problem with this algorithm is that it expects as input a map that does not change over time. This requirement cannot be upheld in the domestic environment, where some furniture is frequently moved (e.g., chairs, small tables, etc.) and new objects can clutter the environment semipermanently.

In order to overcome this limitation, we employ an algorithm which combines PRM with Growing Neural Gas (GNG) [6]. GNG is a neural network with unsupervised learning, used to reduce the dimensionality of the input space. In this kind of network, nodes represent symbols and edges represent semantic connections between them; the Hebbian learning rule is used in many approaches to update nodes and create edges between them. Given a system which has a finite set of outputs, applying the Hebbian rule allows for modifying the network in order to strengthen the output in response to the input. Otherwise, given two outputs that are correlated to a given input, it is used to strengthen their correlation. For our concerns, the nodes (symbols) represent locations and the edges the possibility to go from one location to another. In this sense, we can use, together with the Hebbian learning rule, a simple visibility check in order to create a link between two nodes, as PRM does. GNG cannot be straightforwardly used in a robot motion problem, because the topological information is valid only when the graph has reached a state of equilibrium.

5.3 The Interaction Manager

Interaction within RoboCare is a multifaceted problem that presents many interesting challenges. All the agents operating within the environment contribute to form the assistive behavior by maintaining a continuous exchange of updated information among one another. Recent psychological studies performed within the context of the RoboCare project [7] have stressed the importance of using an embodied assistant, the robot, as the main interactor between the environment and the user. Since the robot plays the role of cognitive mediator, "speech" was chosen as the main user-interaction modality.

Verbal interaction to/from the user is enacted by the *Interaction Manager* (IM), which is in charge of controlling and integrating the services exported by the following subsystems:

- The *Speech Synthesis* Module, called Lucia, developed at the Institute of Cognitive Sciences and Technologies of Padua;
- The *Speech Recognition* Module, called Sonic, developed at the University of Colorado.

The Lucia subsystem allows the generation of speech acts according to a text-to-speech fashion; it accepts text strings in input, which are afterward verbally pronounced by a talking face through realistic labial movements. The IM is in charge of assembling the correct textual string to be submitted to Lucia, in conjunction with the ADL monitor constraint analyzer and in accordance to the environment overall status, in order to provide "user-oriented" suggestions, warnings and answers.

Originally designed for the English language, the Sonic subsystem has been enhanced for the Italian language during the third year of the RoboCare project's development, through the implementation of an Italian grammar that allows to capture a basic lexicon based on Italian phonemes. The recognized words and/or sentences

are successively returned in text format; the IM is in charge of interpreting the returned strings and adjusting the involved environmental variables in accordance to
the received messages. Both the previous modules have been lightly reengineered
in order to suite the ROBOCARE environment's utilization needs.

In summary, the interaction between the intelligent assistant and the user may
occur in two different directions:

User → Intelligent Assistant: the assisted person can interact with the assistant
according to two different modalities, one of which is verbal communication. General questions can be asked to the assistant, which eventually provides the answer
(e.g., *"At what time should I take my pill?"*). It is worth highlighting how this type
of interaction can be considered as "passive" from the intelligent assistant point of
view, because the system only reacts if some sort of action is requested by the user
(*on demand* interaction). Alternatively, interaction is ensured through the use of
an ordinary Personal Data Assistant (PDA) endowed with a user-friendly interface
which allows the user to have a direct control on the Robot Motion Skills. Through
the PDA it is possible to impart commands to the robot and order it to move to a
specific location, to relay streaming video from the robot's current location to the
PDA, as well as to stop the robot.

Intelligent Assistant → User: in order to provide a truly proactive environment,
an intelligent assistant should also be able to autonomously understand when to
interact, in order to support the user. For this reason our work also focused on
the implementation of active services from the assistive environment. According
to the ROBOCARE view, the activities that are to be monitored describe the behavior that the assisted person should adhere to; such activities cannot be mandatory,
even though their execution represents an important objective for the safeguard of
the assisted person's health. Therefore, failing to perform some prescribed action
is considered as a *trigger* by the system to start a dialog with the user. In general,
the violation of the constraints that exist among the activities is intented as a trigger
for the system to take the initiative and perform some actions, like approaching the
assisted person to offer assistance, or issuing verbal warnings and/or suggestions
(*spontaneous* interaction).

5.4 Environmental Sensors for People Tracking and Posture Recognition

A major objective of the ROBOCARE project was the integration of different intelligent components that are deployed not only on board a mobile robot, but also as
"intelligent" sensors in the environment. In particular, we have developed a People Localization and Tracking service[4] (PLT) based on a stereo vision sensor, which
provides the means to locate the assisted person and other people in the environment.
This environmental sensor was deployed at RoboCup@Home 2006 in Bremen in the

---

[4] http://www.dis.uniroma1.it/~iocchi/PLT

form of an "intelligent coat-hanger", demonstrating easy setup and general applicability of vision-based systems for indoor applications. The system is scalable as multiple cameras can be used to improve area coverage and precision. In addition, vision-based posture recognition can be cascaded to the PLT computation in order to provide further information on what the assisted person is doing.

Our stereo-vision-based tracking system is composed of three fundamental modules: (1) background modeling, background subtraction and foreground segmentation, which are used to detect foreground people and objects to be tracked; (2) plan-view analysis, which is used to refine foreground segmentation and to compute observations for tracking; (3) tracking, which tracks observations over time maintaining association between tracks and tracked people (or objects).

The PLT service is effectively capable of tracking the position of a human being within a domestic environment. In addition, the system is resilient to changes in the lighting conditions of the environment, thus enabling portability and easy setup (as demonstrated at the RoboCup@Home competition). This characteristic is particularly useful in domestic environments, where strong differences may occur due to artificial and natural lighting conditions. The key solutions which have made these features possible are:

1. The background model, which is a composition of intensity, disparity and edge information; it uses a *learning factor* that varies over time and is different for each pixel in order to adaptively and selectively update the model; moreover, it uses a new notion of *activity* based on edge variations.
2. Plan-view projection computes *height maps*, which are used to detect people in the environment and refine foreground segmentation in case of partial occlusions.
3. *Plan-view positions* and *appearance models* are integrated in the tracker and an optimization problem is solved in order to determine the best matching between the observations and the current status of the tracker.

The output of these three phases of the computation is depicted in Figure 5.2.

In addition to the PLT service, the system also provides a Posture Recognition (PR) service. Specifically, this module is cascaded to the PLT module, as its input is the person-blob obtained by the PLT algorithm. In addition, the service relies on a 3D human body model which has been carefully chosen by considering the quality of data available from the segmentation steps. In our application the input data are not sufficient to cope with hand and arm movement. This is because arms are often missed by the segmentation process, and noises may appear as arms. Without taking into account arms and hands in the model, it is not possible to retrieve information about hand gestures, but it is still possible to detect most of the information that allows to distinguish among the principal postures, such as *STANDING*, *SITTING*, *BENT*, *KNEELING*, and *LAYING*. Our application is mainly interested in classifying these main postures and thus we adopted a model that does not contain explicitly arms and hands.

A detailed description of the PLT and PR services is outside the scope of this paper, and the interested reader is referred to [8, 9] for further descriptions of the technology underlying the PLT and PR services. Nevertheless, we should underscore

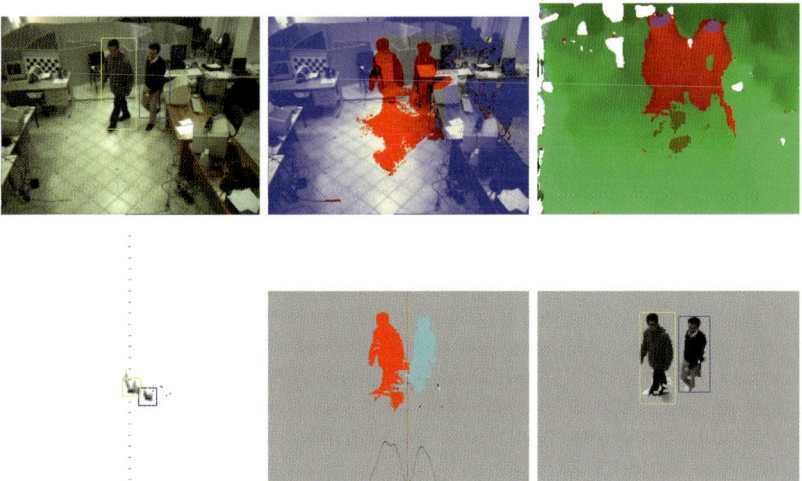

Fig. 5.2 The phases of the PLT service (from left to right, top to bottom): original image, intensity foreground, disparity foreground, plan-view, foreground segmentation, and person segmentation.

that these services are key enabling factors for the sophisticated cognitive support services provided by the smart home. Constant tracking and posture recognition allows to deduce the state of the assisted person, and is therefore responsible for activity recognition. As we briefly explain in the next Sections, recognized activities are propagated within a temporal representation of the assisted person's daily schedule, which in turn triggers the proactive behavior of the robotic mediator (in the form of suggestions, warnings, and so on).

5.5 Monitoring Activities of Daily Living

This Section focuses on the Execution Monitoring System, or Activity of Daily Living (ADL) monitor. Activity recognition and management plays a significant role in advice and warning synthesis, as will be shown shortly. The desired behavior is initially synthesized in terms of a set of activities to be monitored (schedule), bound to one another through complex temporal relationships. These temporal links are of great importance: in fact, not only does the schedule need to be constantly monitored in order to know *which* activities are indeed being executed; also, the time at which the activities are performed is essential, as delays or anticipations on temporally related tasks might trigger some initiative on behalf of the monitoring system. Through temporal constraint analysis, the ADL monitor decides which pieces of information to store and make available to the other agents, in order to ensure a correct global reaction. Some intervention might even be directly triggered by the ADL monitor analysis itself in a more reactive fashion, depending on the gravity of the occurred circumstance. In general, the system is able to assess the situation by querying all available agents, which are designed to act independently and asynchronously.

5.5.1 Schedule Representation and Execution Monitoring

The scheduling technology underlying the whole system is based on Constraint Satisfaction Problem (CSP) solving techniques. More specifically, the baseline schedule defined by the caregiver (a doctor, or a family member) is represented in a temporal CSP, usually called Temporal Constraint Network (TCN) [10]. The variables in a temporal CSP represent the time points, which can be constrained one another by binding the distance between any two variables. Every activity in the schedule is associated with two time points (the start and the end time); by imposing distance constraints among the time points in the TCN, it is possible to define complex temporal relations among the activities, task durations as well as general separation constraints. TCN's consistency on insertion of new time points and/or new constraints among existing time points is checked through proper propagation algorithms [10].

In our schedule representation model, each activity T_i is characterized by a start point $st(T_i)$, an end point $et(T_i)$ and a duration $d(T_i)$. A Schedule is defined as a set of such activities that are supposed to be bound by *minimum* and/or *maximum* temporal constraints imposed by the caregiver: the duration of the task T_i can be set by defining temporal constraints between $st(T_i)$ and $et(T_i)$, while by properly constraining the start and end times of different tasks, it is generally possible to model the desired temporal relationships among T_i and any other activity of the schedule.

Under the hypothesis that the schedule represents the activities that have to be monitored, each imposed constraint helps to specify the desired behavior we would like the assisted person to adhere to. For instance, activity T_i might represent the activity of having breakfast and T_j the activity of taking a medicine: in this case, it is possible to impose constraints so as to model the circumstance that the medicines should not be taken neither too soon nor too late after eating; the values associated with the temporal constraints quantitatively specify the extent of such bounds. According to this representation it is possible to describe behavioral patterns which can be very complex, either in terms of the number of activities involved, or in terms of temporal constraints which may insist among them.

Note that this modeling paradigm implicitly allows for temporal flexibility in the synthesis of the desired behavioral pattern: in fact, the possibility to introduce minimum and maximum temporal constraints permits to specify temporal slacks in order to allow for some tolerance before a constraint is deemed violated. It is straightforward how this is the only viable solution in the context of execution monitoring of human behaviors, as it avoids putting the assisted person (and the caregiver!) against unacceptably strict, and thus unmanageable, action sequences.

The problem of execution monitoring of activities belonging to a predefined schedule represents a delicate issue, the main reason being that the words "control" and "monitoring" are often interpreted as synonyms[5]. For precision's sake, "control" should be interpreted as the deployment of a corrective action aiming at altering the state of the world, while "monitoring" should be simply interpreted as the action of observing reality, giving up any volition of interference. Therefore, scheduling can play a different role, depending on the particular domain: the more the domain allows to see the schedule activities as commands to be dispatched in the environment, the more that scheduling can be seen as a *control action*, as the sequencing decisions on the activities will directly influence the future evolution of the world.

In the ROBOCARE context, we obviously have no control whatsoever in the actions the assisted person is going to perform, despite the caregiver's prescriptions. Therefore, the task of following the person's behavior falls exclusively in the *monitoring* category. The system limits itself in taking note of the evolution of the environment, continuously keeping an updated internal representation of the latter, and possibly reacting to some significant events, if deemed necessary. The monitoring efforts will therefore focus upon: (1) keeping the internal representation of the real world consistent with the behavioral decisions of the assisted person at all times, and (2) performing the necessary rescheduling actions so as to keep at a maximum the number of temporal constraints originally imposed by the caregiver. This second point is of great importance as the maintenance of temporal information in terms of constraints is essential in order to perform correct situation assessment and/or future-consistent *what-if* analysis.

5.5.2 Constraint Management in the ROBOCARE Context

An extremely important role in the execution monitoring problem within the ROBO-CARE context is played by the management of all the temporal constraints present in the schedule. As the environment sensing cycle commences, the system periodically checks the state of the monitored area, trying to detect and recognize the execution state of all the activities.

Regardless of the prescribed behavior represented in the baseline schedule, the assisted person is obviously free to act as she likes: this basically means that at each detection cycle, the system is called to precisely assess the possible differences between the actual and desired state. Assessing such differences does not necessarily entail the necessity for a system reaction, as the schedule is in general synthesized according to flexibility criteria: only when a true constraint violation occurs, shall reaction be triggered.

To be more concrete, let us consider the monitoring of a behavioral pattern described by a schedule composed of activities $A = \{a_1, a_2, \ldots, a_n\}$ be the set of

---

[5] In the Italian language, for instance, "to control" and "to monitor" are translated with the same term "*controllare*".

activities involved, and $C = \{c_1, c_2, \ldots, c_m\}$ the set of temporal constraints imposed among the activities. In order to represent an *executable* schedule, $\langle A, C \rangle$ must be both temporally and resource consistent. It is the responsibility of the caregiver to synthesize a *semantically* correct plan, while the system is able to detect possible temporal and resource inconsistencies, after the problem loading phase. In case of resource inconsistencies (i.e., should the assisted person be wrongly scheduled to perform two activities at the same time), the system automatically proposes an alternative plan and waits for the caregiver's acceptance; instead, temporal inconsistencies require immediate corrective intervention on behalf of the user.

Algorithm 5.1 The Execution Monitoring Algorithm.

---

1.   **while** true **do**
2. $Events_t \leftarrow S_t$
3.   **if** $Events_t \neq \emptyset$ **then**
4. $C_{r,t} \leftarrow$ removeConstraints()
5.   insertContingencies($Events_t$)
6. $K_t \leftarrow \emptyset$
7.   **while** $C_{r,t} \neq \emptyset$ **do**
8. $c_j \leftarrow$ chooseConstraint($C_{r,t}$)
9.   **if** $\neg$ re-insertConstraint(c_j) **then**
10. $K_t \leftarrow K_t \cup c_j$

Algorithm 5.1 shows the execution monitoring algorithm employed in the ROBO-CARE context. As shown in the algorithm, an "environment sensing" action is periodically performed (line 2). This occurs by accessing the symbolic representation of the current situation (S_t). As we show in Section 5.6, this information is obtained by means of a cooperative multiagent deduction process. The details of how deduction occurs starting from the symbolic information deriving from the sensors are the object of Section 5.6. As a result, the set $Events_t$ of the occurred events is periodically acquired. By *event* we mean any mismatch between the expected situation, according to the caregiver's prescriptions, and the actual situation (i.e., a planned action which fails to be executed, is considered as an event).

If events are detected, the first action is to remove all the active constraints present in the schedule (line 4). By *active* constraints, we mean those which do not completely belong to the past, with respect to the actual time of execution t_E. More formally, given an execution instant t_E and a constraint c_k binding two time points t_a and t_b, c_k is considered *idle* if and only if $(t_a < t_E) \wedge (t_b < t_E)$. All constraints that are not idle are active. Obviously, idle constraints do not take part in the analysis because they will not play any role in the evolution of the future states of the world.

In the next step (line 5) all the detected contingencies, properly modeled as further constraints, are inserted in the plan. This is the step where the system updates the internal representation of the schedule in order to preserve consistency with the world's true state.

Lines 7–10 implement the constraint reinsertion cycle, where the algorithm tries to restore as many caregiver requirements as possible given the current situation. Notice in fact that it is probable that not all the original constraints will be accepted at this point: the occurrence of the contingencies might in fact have changed the temporal network constrainedness, so as to make impossible the complete reinsertion of the constraints removed at the previous step. During the cycle, all the constraints which are rejected are stored in the set K_t.

Constraint insertion (and rejection) is an extremely delicate issue, for at least three reasons. First, system reaction may consist in verbal suggestions or warnings: the information conveyed by these messages strongly depends on the contents of the set K_t. The analysis of all the rejected constraints quantitatively and qualitatively determines the system's response. Given a temporal network TN underlying the current schedule, the set $K_t = \{k_{t,1}, k_{t,2}, ..., k_{t,r}\}$ must be such that: (1) the insertion of each $k_{t,j}$ in TN causes a propagation failure; (2) the cardinality of K_t is maximum. Condition (1) ensures that every constraint in K_t plays a role in determining the system's reaction, ruling out false-positive situations; condition (2) ensures that no contingency escapes the system's attention.

Second, the acceptance of each constraint c_j (and complementarity, the contents of K_t) is generally dependent on the particular order chosen for reinsertion. In general, a number of different choice heuristics (chooseConstraint() method) can be envisaged, leading to different approaches for contingency management. To clarify this issue, let us consider a temporal network TN and two constraints c_1 and c_2 such that the attempt of posting both of them in TN would determine an inconsistency: in this case, if the insertion order is $\{c_1, c_2\}$, then c_2 is going to be rejected; if the opposite order is used, c_1 is rejected. Since in the ROBOCARE context it is essential that the reaction be related to the closest contingency with respect to execution time t_E, the particular heuristic employed for reinsertion is backward-chronological. The result of this choice is that the rejected constraints will be the ones which are temporally closer to the actual instant of execution, therefore meeting the condition of reaction urgency. In other terms, the ROBOCARE monitoring system is oriented toward synthesizing a suggestion regarding the primary cause of a violation, rather than forming one based on a distant effect of the assisted person's behavior. The constraints are chronologically ordered taking into account the values of the time point pairs they are connected to. More formally, given a set of constraints $\{c_1(t_{1,s}, t_{1,e}), c_2(t_{2,s}, t_{2,e}), ..., c_n(t_{n,s}, t_{n,e})\}$, where each $c_i(t_{i,s}, t_{i,e})$ connects the time points $t_{i,s}$ and $t_{i,e}$, the constraint $c_i(t_{i,s}, t_{i,e})$ chronologically precedes the constraint $c_j(t_{j,s}, t_{j,e})$, if $min(t_{i,s}, t_{i,e}) < min(t_{j,s}, t_{j,e})$.

Lastly, the importance of maximizing the number of accepted constraints is directly linked to the need to maintain a schedule's representation which is at all times as close as possible to the original specifications, despite the assisted person's actions. The reason is twofold:

1. The system should at all times be able to give correct answers to questions related to future allocations of the activities, as well as to the temporal bounds imposed among them: Questions like: "At what time do I have to take my medication?"

or "How much time have I got between lunch and dinner?" should always be answered correctly (according to the original caregiver's plan);
2. The system should retain the ability to perform correct *what-if* analysis, in order to deliver reliable information in case of requests like:"If I go for a walk at four o'clock, will I come back in time to watch my favorite TV show?" It is straightforward how the reliability of the answer is strictly related to the quantity of original temporal information that the system is able to retain during the monitoring.

5.5.3 From Constraint Violations to Verbal Interaction

Objective of this Section is to give a hint of how the information regarding the constraint violations can be interpreted by the Interaction Manager (see Section 5.3) into semantically meaningful speech acts that the user may immediately understand.

As we have seen in the description of Algorithm 5.1, each element in the violated constraints set K_t is either a *minimum* or a *maximum* constraint. Temporal relations between activities are generally represented through minimum and/or maximum constraints imposed between the end time of the previous activity and the start time of the following activity; *duration* constraints are defined through both a minimum and a maximum constraint insisting between the start and end time points of the same activity and representing, respectively, the minimum and the maximum duration allowed.

At a basic level, the violation of each constraint can immediately be given the following semantic interpretation, depicted in Figure 5.3:

- the violation of the minimum constraint c_{min}^{ij} between activities A_i and A_j (where A_i can be the *SOURCE* activity[6]), directly involves the following semantics: "A_j *is taking place too soon.*" (Figure 5.3(a));
- the violation of the maximum constraint c_{max}^{ji} between activities A_j and A_i (where A_i can be the *SOURCE* activity), involves the semantics "A_j *is being delayed too much.*" (Figure 5.3(b));

Duration constraints undergo a slightly different analysis: in fact, a violation of a duration constraint on activity A_i might either entail the violation of the minimum or of the maximum constraints involved:

- the violation of the minimum duration constraint implies the semantics: "A_i *has lasted too short.*" (Figure 5.3(c));
- the violation of the maximum duration constraint implies the semantics: "A_i *is lasting too long.*" (Figure 5.3(d)).

The previous bullets represent the *building blocks* for higher level interpretations of the events related to constraint violations. Through a deeper analysis of the temporal

---

[6] The *SOURCE* is a particular activity with zero duration whose start time coincides with the origin of the temporal axis.

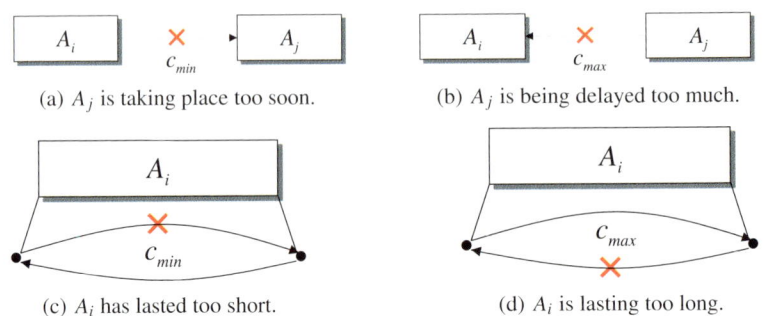

(a) A_j is taking place too soon.   (b) A_j is being delayed too much.

(c) A_i has lasted too short.   (d) A_i is lasting too long.

Fig. 5.3 The Building Blocks for Speech Act Generation based on constraint violations.

network underlying the assisted person's monitored behavior, it is possible to capture a number of useful interrelations among the violated constraints: the idea is to build more and more articulated responses through the fusion of all the semantically useful data carried by each building block associated to a violation.

Nonetheless, constraint violation alone is generally not enough to synthesize meaningful warning speech acts; integrating in the analysis the execution state of the scheduled activities provides a great deal of meaningful information. This semantic integration is performed by the IM agent, which is also responsible for the coordination and management of information synthesis and exchange to/from the robot, through the talking face.

As an example, let us consider a simple plan consisting of a *cooking* and a *lunch* activity, where the *cooking* must be executed before *lunch*, and the two activities must be separated by a temporal constraint of at least 30 minutes (in other words, the assisted person is expected to cook first, then wait at least 30 minutes, and finally have lunch). During the monitoring process, if the system detects a violation of the minimum constraint existing between the two activities, depending on whether the assisted person has performed the cooking or not, the system might respectively suggest to delay the lunch or to prepare something warm to eat.

5.6 Multiagent Coordination Infrastructure

Coordination of multiple services is achieved by solving a Multiagent Coordination (MAC) problem. The MAC is cast as a Distributed Constraint Optimization Problem (DCOP), and solved by ADOPT-N [3], an extension of the ADOPT (Asynchronous Distributed Optimization) algorithm [11] for dealing with n-ary constraints.

One of the most crucial issues which arise when integrating diverse agents is that of coordination. Specifically, the combination of basic services provided by all these agents is accomplished by a distributed constraint reasoning infrastructure. The coordination scheme provides a "functional cohesive" for the elementary services, as it defines the rules according to which the services are triggered. Each service corresponds to a software agent to which tasks are dynamically allocated in function of the current state of the environment and of the assisted person. For instance, if the PLT and PR services recognize that the assisted person is lying on the floor in the kitchen (a situation which is defined as "anomalous" in the overall rule set), then the coordination mechanism will trigger the robot to navigate toward the assisted person's location and ask whether everything is all right.

The coordination of the above-mentioned elementary services is defined so as to demonstrate complex added value services which require the cooperation of multiple elementary services. Some examples of global behaviors are the following:

Scenario 1 *The assisted person is in an abnormal posture-location state (e.g., lying down in the kitchen).* **System behavior:** *the robot navigates to the person's location, asks if all is well, and enacts a predefined contingency plan, such as placing an emergency phone call.*

Scenario 2 *The ADL monitor detects that the time bounds within which to take a medication are jeopardized by an unusual activity pattern (e.g., the assisted person starts to have lunch very late in the afternoon).* **System behavior (option 1):** *the robot will reach the person and verbally alert him/her of the possible future inconsistency.* **System behavior (option 2):** *the inconsistency is signaled through the PDA.*

Scenario 3 *The assisted person asks the robot, through the PDA or verbally, to go and "see if the window is open".* **System behavior:** *the robot will navigate to the designated window (upon obtaining its location from the fixed stereo cameras) and* **(option 1)** *relay a streaming video or snapshot of the window on the PDA, or* **(option 2)** *take a video/snapshot of the window, return to the assisted person and display the information on its screen.*

Scenario 4 *The assisted person asks the intelligent environment (through the PDA or verbally to the robot) whether he/she should take a walk now or wait till after dinner.* **System behavior:** *the request is forwarded to the ADL monitor, which in turn propagates the two scenarios (walk now or walk after dinner) in its temporal representation of the daily schedule. The result of this deduction is relayed to the assisted person through the PDA or verbally (e.g., "if you take a walk now, you will not be able to start dinner before 10:00 pm, and this is in contrast with a medication constraint").*

5.6.1 Casting the MAC Problem to DCOP

As mentioned, multiagent coordination is cast as a distributed constraint optimization problem and solved by the agents according to the (distributed) ADOPT-N algorithm. Specifically, a distributed constraint optimization problem is a tuple $\langle \mathcal{V}, \mathcal{D}, \mathcal{C} \rangle$ where $\mathcal{V} = \{v_1, \ldots, v_n\}$ are variables with values in the domains

$\{D_1, \dots, D_n\} = \mathcal{D}$, and $\mathcal{C}$ is a set of constraints among variables. Constraints may involve an arbitrary subset of the variables (n-ary constraints): a constraint among the set $C \subset \mathcal{V}$ of k variables is expressed as a value function in the form $f_C : D_1 \times \dots \times D_k \to \mathbb{N}$. For instance, a constraint involving the three variables $\{v_1, v_3, v_7\}$ may prescribe that the cost of a particular assignment of values to these variables amounts to c, e.g., $f_{v_1, v_3, v_7}(0, 3, 1) = c$. The objective of a constraint optimization algorithm is to calculate an assignment $\mathcal{A}$ of values to variables while minimizing the cost of the assignment $\sum_{C \in \mathcal{C}} f_C(\mathcal{A})$, where each f_C is of arity $|C|$.

In the specific case of the RDE, the cost function is modeled so as to reflect the desiderata of system behavior. Specifically, the domains of the variables model the states of the services (i.e., what the system can provide) as well as the possible states of the environment and of the assisted person (i.e., what can occur). Constraints bind these variables to model relations among services, that is, the overall behavior of the smart home and how knowledge is shared among the agents. A high-level representation of how the RDE's components are connected to the underlying DCOP problem formulation is shown in Figure 5.4.

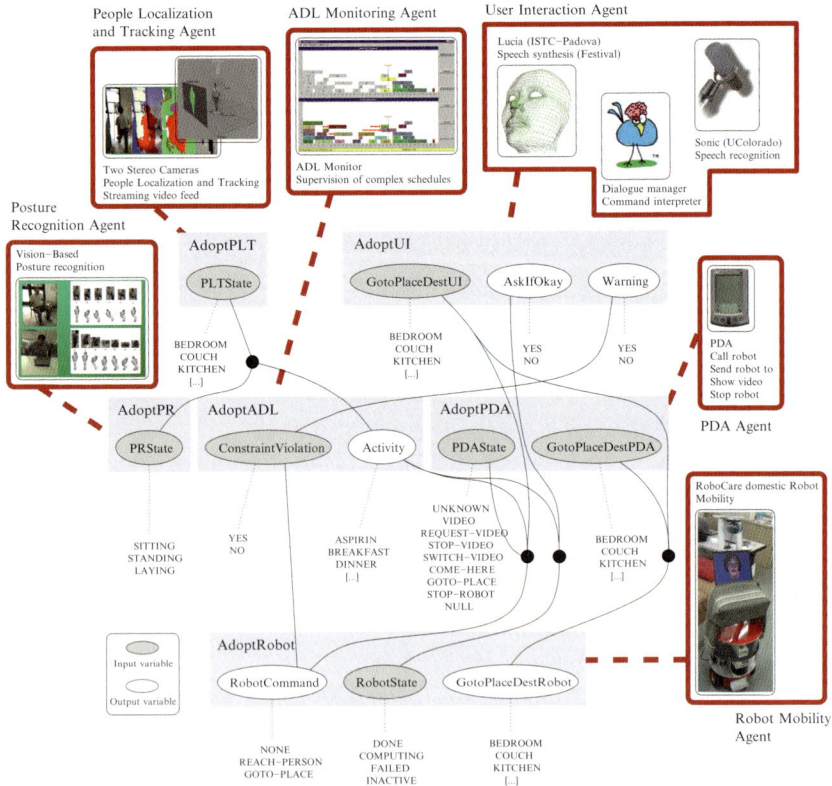

Fig. 5.4 The ROBOCARE DCOP for service integration.

In the RDE DCOP, variables represent input for the decision process and/or instructions for controlling the enactment of the services provided by the RDE. For instance, the *PLTState* variable represents the position of the assisted person in the environment (whose domain is the enumeration {KITCHEN, LIVINGROOM, BATHROOM, BEDROOM, UNKNOWN}, while the *PRState* variable carries the information on the person's posture (the domains of the variables are shown in the figure). These two variables are purely "sensory", as their value is determined by the sensory input obtained from the PLT and PR services. An example of "enactment" variable is *RobotCommand*, which is set autonomously by its agent according to the decisions taken during the execution of the ADOPT-N cooperative solving algorithm. Moreover, agents can have more than one variable. This is the case of the agent representing the robot, which also has the variable *RobotState* representing the current task in which the robot is engaged, i.e., {DONE, COMPUTING, FAILED, INACTIVE}.

The value functions which model the constraints in the system describe a global cost function whose minima represent the desired system behavior. In first approximation, it is easy to see this reduction as a set of crisp constraints: all consistent states evaluate to a global cost of 0, while inconsistent situations evaluate to ∞. Consistent states establish a correspondence between observations from the sensors and the desired combination of behaviors of the services. For reasons of space we cannot describe the full set of constraints which models the behavior of the RDE as it is instantiated in the ROBOCARE lab. One meaningful example of such constraints is the following: when the PLT and PR sensors assess that there is an emergency situation (e.g., the assisted person is lying on the floor in the kitchen), the *PLTState* and *PRState* variables are set to *KITCHEN* and *LYING*, respectively; we wish to model the fact that the variable representing the assisted person's current activity (*Activity*) should be set to *EMERGENCY* in the event of anomalous situations such as this one.

Following the same reasoning, we can model the overall activity recognition problem in the RDE with a ternary constraint $f_C = \{PLTState, PRState, Activity\}$, depicted in Figure 5.5.

In addition, we add tuples stating that all assignments that are not described by the above zero-cost tuples (e.g., standing in the bathroom having lunch) have infinite cost. This operation can be done automatically, as it consists of calculating the transitive closure of the zero-cost tuples, and associating to the newly generated tuples infinite cost.

PRState	PLTState	Activity	f_C
SITTING	KITCHEN	LUNCH	0
STANDING	KITCHEN	COOKING	0
LAYING	KITCHEN	EMERGENCY	0
STANDING	BATHROOM	MEDICATION	0
LAYING	BATHROOM	EMERGENCY	0
LAYING	BEDROOM	NAP	0
SITTING	≠ KITCHEN	UNKNOWN	0
STANDING	≠ KITCHEN, ≠ BATHROOM	UNKNOWN	0

Fig. 5.5 Constraint modeling activity recognition through the PLT and PR services.

Constraints such as the above are employed to model all aspects of the RDE, such as the proper enactment of the robotic mediator in case of an emergency (i.e., when the Activity variable indicates that the assisted person, is in a state of emergency, the robot will enact a predefined plan as described in scenario 1).

5.6.2 Cooperatively Solving the MAC Problem

As noted, an ADOPT-N agent is instantiated for each service provided by the components of the RDE. Given the current situation S, these agents communicate to each other messages which allow them to trigger the appropriate behavior. Clearly, the state of the environment, of the assisted person, and of the services themselves changes in time: let the situation (i.e., the state of the environment, of the assisted person and of the services) at time t be S_t. The DCOP formulation of the MAC described earlier represents the desired behavior of the system in function of the possible states of the RDE. Therefore, if $S_t \neq S_{t-1}$, the ADOPT-N agents must trigger an "instance of coordination" so as to decide the assignment $\mathscr{A}$ which represents the desired enactment of services.

One of the challenges of the RDE scenario with respect to distributed coordination is the heterogeneity of the agents. The strong difference in nature between the various components of the RDE reflects heavily on the coordination mechanism. This is because of the uncertainty connected to the time employed by services to update the symbolic information which is passed on to the agents.

Algorithm 5.2 Synchronization schema followed by each ADOPT-N agent a in the RDE.

```
t ← 0
S_t ← getSensoryInput(V_a)
while true do
    S_{t-1} ← S_t
    while (S_t = S_{t-1}) ∧ (t ≥ a'.t, ∀a' ≠ a) do
        S_t ← getSensoryInput(V_a)
    t ← t + 1
    forall d_i ∈ D_{v∈V_a} do
        lb(d_i) ← 0        /** Reset lower and **/
        ub(d_i) ← ∞       /** upper bounds   **/
    𝒜|_{V_a} ← runAdopt()  /** Iteration terminates on ADOPT-N termination **/
    triggerBehavior(𝒜|_{V_a})
```

As a consequence, it is in general impossible to have strict guarantees on the responsiveness of the agents. For this reason the albeit asynchronous solving procedure needs to be iterated synchronously. More specifically, ADOPT-N is deployed in the RDE as described in Algorithm 5.2, according to which the agents continuously monitor the current situation, and execute the ADOPT-N algorithm whenever a difference with the previous situation is found. The getSensoryInput() method in the pseudo-code samples the state of the environment which is represented by agent a's variables V_a (what we have informally called "sensory" variables). Specifically, the values of these variables are constrained to remain fixed on the sensed value during the execution of the ADOPT-N decision process. In practice, this occurs by posting a unary constraint which prescribes that any value assignment which is different from the sensed value should evaluate to ∞, and is therefore never explored by the agent controlling the variable. This constraint posting mechanism is a feature of ADOPT-N. Clearly, it is also possible to restrict the values of these variables by modifying the problem before each iteration. The constraint posting strategy was employed to facilitate representation and reuse of code. In fact, the DCOP problem never needs to change between iterations, and this allows to minimize the reinitialization phase between iterations (which can be reduced to resetting the lower and upper bounds of the domain values for each variable as shown in the algorithm — see [3, 11] for details on the ADOPT and ADOPT-N algorithms). Moreover, posting a unary constraint on a variable for the entire duration of the solving process does not affect the computational complexity of the algorithm.

Notice, though, that ADOPT and its variant ADOPT-N do not rely on synchronous communication between agents, thus natively supporting message transfer with random (but finite) delay. This made it possible to employ ADOPT-N within the RDE scenario without modifying the algorithm internally. Furthermore, while most distributed reasoning algorithms (like ADOPT itself) are employed in practice as concurrent threads on a single machine (a situation in which network reliability is rather high), the asynchronous quality of ADOPT-N strongly facilitated the step toward "real" distribution, where delays in message passing increase in magnitude as well as randomness.

5.7 Conclusions

During the first two years of project development, efforts were concentrated on developing the technology to realize the individual components (or services) of the RDE. The services provided by this technology were deployed in the environment according to a service-oriented infrastructure, which is described in [12]. This allowed to draw some interesting conclusions on the usefulness of robots, smart sensors, and proactive domestic monitoring in general (see, e.g., [13]).

In the final year of the project, and in part toward the goal of participating in the RoboCup@Home competition, the attention shifted from single component development to the functional integration of a continuous and context-aware environment. The issue was to establish a convenient way to describe how the services should be interleaved in function of the feedback obtained from the sensory subsystem and the user. The strategy we chose was to cast this problem, which can also be seen as a service-composition problem, in the form of a multiagent coordination (MAC) problem.

It is interesting to notice that the specific constraint-based formulation of the MAC problem is strongly facilitated by the possibility to encode n-ary constraints. As discussed, this is convenient for modeling the functional relationship among multiple services as it allows to precisely indicate the relationships between sensed input and the resulting enactment. Another advantage of the constraint-based formulation is that the system is easily scalable. In fact, adding another sensor, service or intelligent functionality requires adding an ADOPT-N agent and its variables to the problem, and system behavior can be specified incrementally.

Finally, as noted earlier, an interesting area for future research is the development of more powerful formalisms for specifying service interaction and invocation in terms of a DCOP problem. One of the goals of ROBOCARE has been to develop technology which is at least to a certain degree usable by non-experts[7]. The knowledge acquired in three years of ROBOCARE can certainly contribute to building systems which are close to becoming market-level products.

References

1. F. Pecora, R.R., asconi, G., Cortellessa & A. Cesta, User-Oriented Problem Abstractions in Scheduling, Customization and Reuse in Scheduling Software Architectures, in Innovations in Systems and Software Engineering, **2(1)**, pp.1-16. 2006.
2. A. Cesta, G. Cortellessa, A.Oddi, N. Policella & A. Susi, A Constraint-Based Architecture for Flexible Support to Activity Scheduling, in Proceedings of the 7th Congress of the Italian Association for Artificial Intelligence on Advances in Artificial Intelligence, 369–381, 2001.
3. F. Pecora, P.J. Modi & P. Scerri, Reasoning About and Dynamically Posting n-ary Constraints in ADOPT, in Proceedings of 7th International Workshop on Distributed Constraint Reasoning, 2006.
4. G. Grisetti, C. Stachniss & W. Burgard, Improving Grid-based SLAM with Rao-Blackwellized Particle Filters by Adaptive Proposals and Selective Resampling, in Proceedings of International Conference on Robotics and Automation, 2443–2448, 2005.
5. L. Kavraki & J. Latombe, Probabilistic roadmaps for robot path planning, in Practical Motion Planning in Robotics: Current Approaches and Future Challenges", K.Gupta & A.P. del Pobil editors, Cambridge University Press, 33–53, 1998.
6. B. Fritzke, A growing neural gas network learns topologies, in Advances in Neural Information Processing Systems 7, G. Tesauro and D. S. Touretzky & T. K. Leen editors, MIT Press, 625–632, (1995).
7. M.V., Giuliani, M., Scopelliti & F., Fornara, Elderly people at home: technological help in everyday activities, in Proceedings of the IEEE International Workshop on Robot and Human Interactive Communication, 365–370, 2005.
8. S., Bahadori, L., Iocchi, G.R., Leone, D., Nardi & L., Scozzafava, Real-Time People Localization and Tracking through Fixed Stereo Vision, in Applied Intelligence, **26(2)**, 83–97. 2007.
9. S., Pellegrini, & L., Iocchi, Human Posture Tracking and Classification Through Stereo Vision, in Proceedings of the International Conference on Computer Vision Theory and Applications, 2006.
10. R.Dechter, I.Meiri & J.Pearl, Temporal Constraint Networks, in Artificial Intelligence, **49 (1–3)**, 61–95, (1991).
11. P.J.Modi and W.M.Shen and M.Tambe & M.Yokoo, ADOPT: Asynchronous Distributed Constraint Optimization with Quality Guarantees, in Artificial Intelligence, **161(1–2)**, 149–180, 2005.

---

[7] See, e.g., the behavioral constraint specification formalism used by caregivers described in [1].

12. S.Bahadori, A.Cesta, L.Iocchi, G.R.Leone, D.Nardi, F.Pecora, R.Rasconi & L.Scozzafava, Towards Ambient Intelligence for the Domestic Care of the Elderly, in Ambient Intelligence: A Novel Paradigm, P. Remagnino, G-L. Foresti & T.Ellis editors, Springer 15–38, 2005.
13. A. Cesta & F. Pecora, The RoboCare Project: Intelligent Systems for Elder Care, in Proceedings of the AAAI Fall Symposium on Caring Machines: AI in Elder Care, 25–28, 2005.

Chapter 6
Ubiquitous Stereo Vision for Human Sensing

Ikushi Yoda and Katsuhiko Sakae

Abstract We are now researching real-time recognition technology mainly for humans (of human existence, the face, intentional gestures and trajectories) in the real-world environment. We placed stereo cameras ubiquitously and used both high- and low-speed network. This chapter describes the concept of this "Ubiquitous Stereo Vision" and a new method for human sensing. The objectives of this research were to use 3D information and texture image, and to develop a real-time new human sensing method. We place the stereo cameras statically, utilize range information as a key, and understand the scene of human behavior. The cases of indoor life space support and safety improvement support are described as the specific applications.

6.1 Introduction

Computer and communication technologies are changing rapidly, with higher speeds, lower costs, and larger network capacity. Given this, further progress in image recognition and human interface technology using large numbers of cameras is important.

The final goal of our research is to develop a real-time recognition methodology which can adapt to real scenes involving mainly humans (concerning the presence of people, tracking, face recognition, gestures, and so on). We have conducted experiments on general-purpose human sensing in a ubiquitous environment (with multiple cameras) and with a large-capacity network. This chapter describes the basic concepts and techniques for recognizing people with this type of system.

Ikushi Yoda

National Institute of Advanced Industrial Science and Technology, Tsukuba, 305-8568 Japan, e-mail: i-yoda@aist.go.jp

Katsuhiko Sakae

National Institute of Advanced Industrial Science and Technology, Tsukuba, 305-8568 Japan, e-mail: k.sakaue@aist.go.jp

D. Monekosso et al. (eds.), *Intelligent Environments*, Advanced Information and Knowledge Processing, DOI: 10.1007/978-1-84800-346-0_6,

Learning stereo vision [1] has been developed to achieve a more natural inter-action for the man-machine interface. This is fundamental for the concept of one stereo camera for one user. Here, we expand the mechanism of the vision interface to consist of multiple stereo cameras for multiple users. The ultimate purpose of this research is to develop a new method of learning and recognizing multiple images in-put from a large number of cameras connected via a high-speed network. We utilize range information from the stereo cameras to clip the areas containing the people at a location.

There has been much research on 3D reconstruction using information from a large number of cameras [3, 4, 5]. Using images from multiple viewpoints, much of this research focuses on contents generation. The generation of a middle viewpoint has thus become a realistic goal.

On the other hand, there is also research that simply uses the range information obtained from a stereo camera for human sensing [6]. Stereo cameras are placed in a room, and the system recognizes the position of a person. Not only stand-alone stereo cameras, however, but also touch sensors and fingerprint ID devices are used in this research. Therefore, the vision sensors are mere tools, and the research fo-cuses on more effective utilization of personal computers.

There is also much research on vision systems based on elemental technologies [2]: distribution, cooperation, and activation. Most of this research, however, uses active cameras, which are slower than real people.

In the experiments described here, we installed stereo cameras ubiquitously. Re-gions containing multiple people were locally clipped by using range information from the cameras, and then all information is recognized and integrated in parallel. Figure 6.1 shows the conceptual scheme of this system, called Ubiquitous Stereo Vi-sion. In our system, 3D and texture (color or monochrome) image information are acquired at a video or semi-video rate. Multiple people are recognized after segmen-tation by taking the 3D information as a key. In addition, all results are integrated to understand the human behavior at a location.

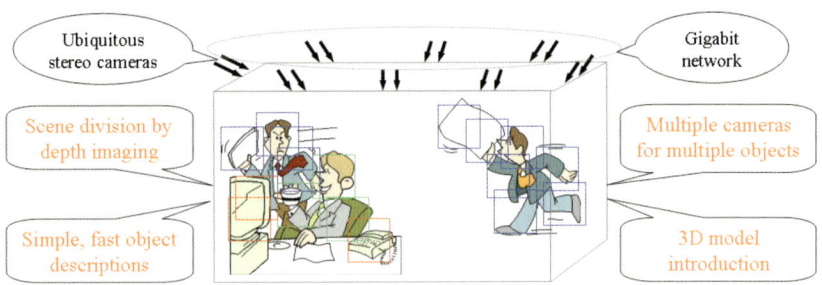

Fig. 6.1 Concept of Ubiquitous Stereo Vision.

6.2 Ubiquitous Stereo Vision

6.2.1 Concept of Ubiquitous Stereo Vision

The Ubiquitous Stereo Vision (USV), which we propose here, conceptually provides real-time operation, in a real environment, with real utilization. USV should be easily adaptable to real environments. Table 6.1 lists key specifications for the concept of USV.

For acquisition of 3D structures, a control device connected with each stereo camera reconstructs 3D information and extracts features as a local process. We call this device a *stereo camera server*. Next, the range information and color or monochrome images from each stereo camera server are sent to a PC via a network. By integrating all the results on one computer, the system acquires 3D information as a global process. We call this computer integrating all necessary information from the stereo camera servers an *application client*. The integration method depends on the application.

6.2.2 Server-Client Model for USV

The stereo camera servers and the application client must work regardless of the number of stereo camera servers. We conducted experiments using four to twelve stereo camera servers.

Because the stereo camera servers are integrated and perfectly synchronized when the cameras are placed around a small space, all information is perfectly

Table 6.1 USV specifications.

3D range and texture image information input from multiple stereo cameras.
Utilization in real time.
Utilization in real environments (indoors and outdoors).
Stand-alone operation of stereo cameras.
Static arrangement: one-time calibration of stereo cameras.
Utilization on both high- and low-speed networks.
Basic, non-dynamic controls (pan, tilt, and zoom).
Attention based on range information, with no specific models.
Recognition based on appearance.

integrated while being synchronized. By integrating information from multiple sources, the system should operate robustly, even if the reliability of a camera decreases for some reason.

On the other hand, a large number of cameras are used when the target is a wider space, such as a train platform. Then, it is not necessary to exchange information among all cameras. In this case, each camera works and processes information independently, and the information integrated in real time becomes symbolic information. Table 6.2 classifies the application of USV according to the integration method and network speed.

6.2.3 Real Utilization Cases

For specific utilization cases, we consider human sensing both indoors and outdoors. Table 6.3 lists various cases.

Table 6.2 Integration and independence in USV.

i. Synchronous integration (via high-speed network)
Both range and color information are integrated, and reliability is improved. e.g., controlling indoor space, monitoring railroad crossing
ii. Synchronous integration (via low-speed network)
Range and color information are processed locally, and only necessary information is integrated. e.g., platform safety management, traffic management
iii. Independent operation
One camera operates in a narrow area. e.g., interface for a narrow area, personal interface for severely disabled person

Table 6.3 USV utilization cases indoors and outdoors.

A. Human sensing and interaction indoors (with personal identification)
Acquisition of personal behavior logs Interface for controlling indoor equipments Amusement applications
B. Human sensing outdoors (without personal identification)
Safety management, including train platform edges, railroad crossing Traffic management

Table 6.4 Relationships between people and stereo cameras.

a. Multiple stereo cameras for multiple people
Specific indoor working space
b. Multiple stereo cameras for undefined multiple people
Public space, traffic management, train platform edge, railroad crossing
c. Multiple (single) stereo cameras for single person
Personal vision interface, home care assistance

Table 6.4 classifies the relationships between the people and stereo cameras.

The utilization cases described in the rest of this paper are included in these classifications. In addition, this system and its concept can be adapted to various other cases of human sensing.

6.3 Hierarchical Utilization of 3D Data and Personal Recognition

This section describes the handling of 3D information, which is common to all applications. The basic algorithm must work regardless of the number of stereo cameras. Common purposes for the system include determining and tracking the locations of people, as well as identifying people and recognizing poses. Whether the system operates indoors or outdoors, 3D information is used to achieve these purposes. In principle, we do not choose a precise model like a robot model [11], but rather a model in which algorithm installation is easy and operation is high speed.

6.3.1 Acquisition of 3D Range Information

The stereo cameras are mounted on ceilings or poles and placed facing the area monitored. The origin of a 3D coordinate system derived from a camera is the center of the camera's view field. As Figure 6.2 shows, the plane parallel to the surface of the camera lens (or the camera surface) is the *XY* plane, and the optical axis extending from the lens is the *Z* axis. The 3D coordinate points for the system are thus expressed as (X, Y, Z). The coordinate system differs from camera to camera, and the statistical processing is easier when the coordinates are converted into an integrated coordinate system for the space inside the crossing, as seen on the left side of Figure 6.2. Thus, a coordinate point (X, Y, Z) in the camera's coordinate system is converted into a coordinate point (x, y, z) in the integrated system, according to the following affine equation, Equation 6.1. Here, a - i and t_x - t_z are parameters for

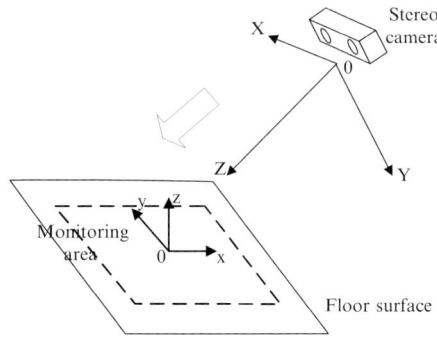

Fig. 6.2 Camera and coordinate systems.

rotation and parallel movement in the affine conversion matrix and are determined by measuring the positions of the cameras.

$$[x\,y\,z\,1] = [X\,Y\,Z\,1] \begin{bmatrix} a & b & c & 0 \\ d & e & f & 0 \\ g & h & i & 0 \\ t_x & t_y & t_z & 1 \end{bmatrix} \begin{vmatrix} a & b & c \\ d & e & f \\ g & h & i \end{vmatrix} \neq 0 \qquad (6.1)$$

6.3.2 Projection to Floor Plane

Searching from acquired range information for the head of a target person can be treated mathematically as a multi-point search problem. Considering the accuracy of the acquired range images and the need for a system working in real time, however, we have converted the problem here to one of image processing.

First, as illustrated in Figure 6.3, we project all points of a range to the floor surface and create eight binary projections P_n at different heights. Here, we refer to these projections as "planes." There are two ways to make the planes. One is the addition draw-out method, which is suitable for representing each person as a cloud while rejecting personal features. The other is the crossing hierarchy draw-out method, which is useful for human poses, because the resulting projections maintain information for each plane level. We can choose either method for the purpose of our application. The 3D coordinate points in space are represented as (x, y, z), while the 2D coordinate points in the projected plane P_n are represented as (X_n, Y_n). Here, P_n is derived by Equation 6.2(1) or 6.2(2), where a is the number of planes, d is the distance between planes, and h is the lowest detected height. The planes are equidistant from one another.

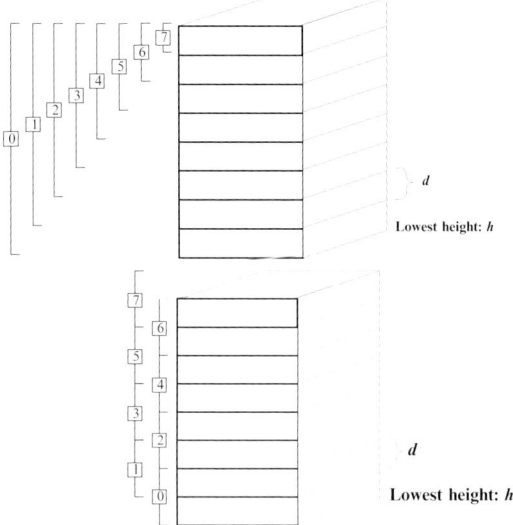

Fig. 6.3 Projections on planes using draw-out methods (with 8 levels). Top: Addition draw-out method; bottom: crossing hierarchy draw-out method.

For the addition method, in order to easily detect a stable center of gravity for the cluster shape of each level, to each plane we add the data from the plane above it.

$$\begin{pmatrix} X_n \\ Y_n \end{pmatrix} = \begin{pmatrix} 1\ 0\ 0 \\ 0\ 1\ 0 \end{pmatrix} \begin{pmatrix} x \\ y \\ z \end{pmatrix} \begin{cases} h + (a-1)d \leq z \leq h + ad & n = a \\ \quad\vdots & \quad\vdots \\ h + d \leq z \leq h + ad, & n = 1 \\ h \leq z \leq h + ad, & n = 0 \end{cases} \qquad 6.2(1)$$

For the crossing hierarchy method, we project the 3D points onto many planes parallel to the floor surface. Because information about the boundary between two planes is lost, the 3D points are projected onto each plane in an overlapping manner, as shown at the bottom of Figure 6.3. Because all small spaces are laid to overlap each other, it is possible to continuously obtain changes in the projected images.

$$\begin{pmatrix} X_n \\ Y_n \end{pmatrix} = \begin{pmatrix} 1\ 0\ 0 \\ 0\ 1\ 0 \end{pmatrix} \begin{pmatrix} x \\ y \\ z \end{pmatrix} \begin{cases} h + (a-1)d \leq z & n = a \\ \quad\vdots & \quad\vdots \\ h + nd \leq z \leq h + (n+2)d, & n = 1 \\ h \leq z \leq h + 2d, & n = 0 \end{cases} \qquad 6.2(2)$$

Figure 6.4 shows an example of an actual projection. The scene includes a man, a girl, and a small child. For the addition method, the shapes of clusters

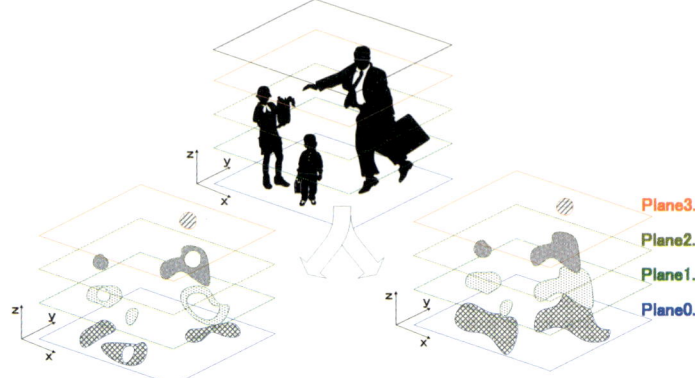

Fig. 6.4 Projection examples (4 levels). Bottom right: addition draw-out method; bottom left: crossing hierarchy draw-out method.

transform gradually between the highest plane and the lowest plane, so we can easily determine the top of each person's head. On the other hand, for the crossing hierarchy method, the outlines of the clusters have significant differences between planes, but we can recognize the plane that includes the man's pointing arm.

6.4 Recognition of Multiple Persons and Interface

There are many systems that detect face positions, face angles, speakers, and so on [7], and commercial face recognition engines [8] also exist. These systems detect faces by using a few cameras and measure face angles, lines of sight, and other characteristics according to detection of structural face elements (eyes, nose, and mouth). Therefore, the recognition object is a human who is viewing a space, such as a specific display, and these systems can be used only in a very narrow space. In addition, the user identification could only coexist with the limited simple detection of characteristics such as face angles, lines of sight, and nodding actions by the systems.

On the other hand, whole-body recognition systems [9] are based on extending the visual field. The resolution of human images in these systems is low. Therefore, they do not provide an interface coexisting with personal identification by gestures or interaction specialized to the person.

In addition, a system of recognizing hand signs [10] utilizes hand images that closed up in a fixed environment in front of a camera, as an interface system. Therefore, this system can work only in a narrower environment. In a wide space like a room, it does not offer a personal interface with hand signs.

Our study aims to use USV to simultaneously identify multiple people and large personal gestures in a room-sized space. This section describes the concepts of personal pose detection and identification, and the use of our system as a personal interface.

6.4.1 Pose Recognition for Multiple People

As an experiment, four stereo cameras were placed in the four ceiling corners of a room (4.5 m × 3.6 m). The functions were to determine the following:

- Who entered the room and when?
- When did they leave?
- What did they do in the room?
- Could they interact with large gestures?

These functions were to be executed in real time.

The goal was thus pose recognition of several people (less than four or five) in a room of about 16 m^2, distinguishing among three conditions: standing, sitting or lying, and movement.

Figure 6.5 illustrates recognition of postures, faces, and an arm-pointing gesture. Two people stand, and one sits. The system recognizes their basic postures and the arm-pointing gesture, and it extracts their faces. Figure 6.6 shows the results of

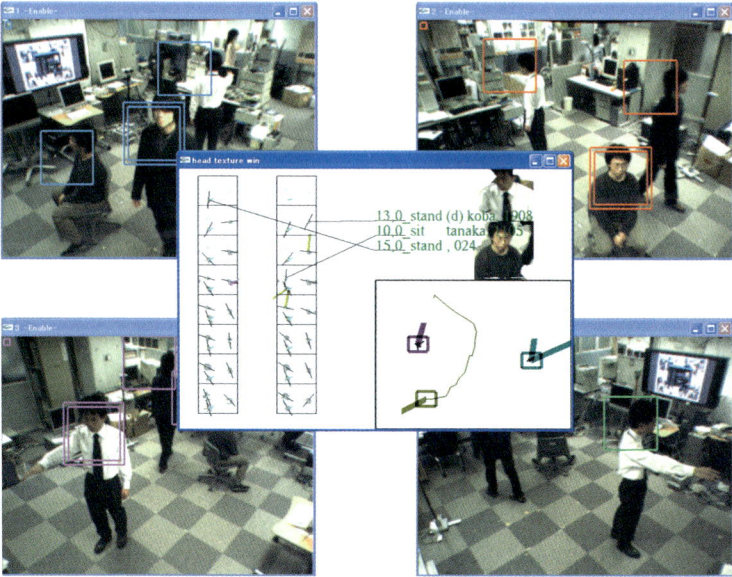

Fig. 6.5 Recognition of postures, faces, and arm pointing.

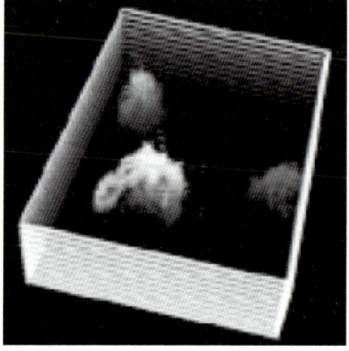 

Fig. 6.6 Utilization of 3D data by the crossing hierarchy draw-out method.

 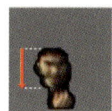  

(a)Input image (b)Mask region (c)Facial region (d)Edge image

Fig. 6.7 Extraction of face images.

projections by the crossing hierarchy draw-out method. The left image of Figure 6.6 shows only the result for 16 planes, while the right image indicates three personal territories and the tops of the three people.

6.4.2 Personal Identification

The personal identification is carried out after personal pose detection and is based mainly on face recognition, although we do not apply normal face recognition from a static, narrow point. Using the range information, we try to recognize selected faces [14]. The direction of the body is obtained from the range information, and the system selects only the front of the face. The results are shown in the center of Figure 6.5. Range information from only one camera was selected, and the face parts were clipped. The system utilized only these faces for learning and recognition. During this learning period, people walked naturally in the room, and the system recognized them under the same conditions. The system automatically classified many face images varying according to the person's position and posture through self-organization (unsupervised learning). A discrimination circuit was then created using only those face images that were suitable for recognition. Figure 6.7 illustrates the process for a clipped face image.

6.4.3 Interface for Space Control

After pose detection and personal identification in the space, the objective is contact-free, unconstrained, position-free interaction. Namely, we aim to develop a personal interface function for the ubiquitous environment by using human sensing technologies.

The initial goal is to recognize arm-pointing gestures by using USV [13]. For example, arm-pointing gestures are used for controlling a mouse cursor in a large display in a specific area. We can expand this function to the whole of an indoor area. Specifically, a large arm-pointing gesture by a person is acquired in real time from a range image. We try to seamlessly recognize the intended arm-pointing gestures in all normal poses (standing and sitting). Furthermore, arm-pointing gestures corresponding to multiple people are useful, except for the case in which there are multiple people at short range (less than about 50 cm). Thus, we aim to create an interface providing personal identification. Possible applications include televisions, air conditioners, curtains, and cable broadcasting. At this time, we utilize Bluetooth for communication and control between the electronic equipment.

6.5 Human Monitoring in Open Space (Safety Management Application)

In this section, we describe experiments with human sensing in huge open spaces, which is different from the above-mentioned case of close human sensing. Specifically, we adapted the system to outdoor spaces. We have conducted many continuous experiments with monitoring railroad crossings, the edges of platforms, and exhibitions.

The purpose of these experiments was to research the possibility of stereo vision in a real situation, in terms of the algorithms, the stereo hardware specifications, and so on. Table 6.5 summarizes the main experiments, which we discuss in turn.

6.5.1 Monitoring Railroad Crossing

Existing sensing technologies for controlling the safety of railroad crossings include the following:

(1)   Electromagnetic sensors consisting of loop coils installed under the road surface, which detect changes in inductance when metal objects pass over the coils; and

Table 6.5 Continuous monitoring experiments for huge open spaces.

	Date	Place	Purpose	Number of cameras
Railroad crossing	2003.1-2	Tokyu Tohyoko Line Motosumiyoshi Railroad Crossing	Monitoring railroad crossing	5 Overlapping and watching obliquely
	2004.6	Tokyu Ooimachi Line Jiyugaoka Railroad Crossing	Monitoring railroad crossing	4 Overlapping and watching obliquely
	2007.2 –	Tokyu Ooimachi Line Jiyugaoka Railroad Crossing	Monitoring railroad crossing (permanent test)	2 Watching obliquely
Platform edge	2001.11	JR Ushiku Station Platform 3	Monitoring platform edge	4 Watching vertically
	2001.11	Tokyu Tohyoko Line Yokohama Station Platform 1	Monitoring platform edge	5 Watching vertically
	2003.1-2	Tokyu Tohyoko Line Motosumiyoshi Station Platform 3	Monitoring wheelchairs and canes on platform	1 Watching obliquely
Huge open space	2005.3-9	Aichi EXPO 2005 Global house Orange hall	Counting audience and analyzing tracks simultaneously with RF-ID technology	7 Watching vertically

(2)   Infrared sensors that detect obstacles by monitoring the blocking of light between emitters and receivers.

These sensors can only detect obstacles of a certain size or larger, and their resolutions are low. They can detect the presence of obstacles but cannot follow their movements, so they are not effective for monitoring people. On the other hand, about 60% of the people killed at railroad crossings are pedestrians in Japan. To detect all people at a crossing and take action flexibly, according to the situation, we developed a stereo vision sensing technology for railroad crossings [15].

The system consists of stereo cameras installed at the corners of a crossing facing the center. The cameras monitor the people passing through the crossing and detect people who are acting dangerously and are within the crossing when a train approaches. By using stereo cameras, the shadows cast by people, trains, and so on, which are problematic for differential-image-based systems, can easily be excluded by extracting range information and using it as a key. Furthermore, by using texture image data along with the range information, it is possible to differentiate birds, scraps of paper blowing around, and other objects that might cause false readings at railroad crossings. Figure 6.8 illustrates the monitoring of a railroad crossing. The four images at the four corners were obtained from four stereo cameras, while the center image is a 3D reconstruction obtained from the four stereo camera servers.

Fig. 6.8 Monitoring a railroad crossing by using USV.

6.5.2 Station Platform Edge Safety Management

Every year, dozens of people die by falling from platform edges at train stations in Japan. In particular, this is an urgent problem for railways in urban areas. USV is utilized for managing the safety of station platform edges [16]. The stereo cameras are arranged in a straight line in order to see the track area from the ceiling of the station. The platform edge and the behaviors of people near the platform edge are monitored. The system distinguishes the following: people, baggage, garbage, birds, and so forth. The purpose is to distinguish whether people and things are in a dangerous place at the platform edge, and whether a human has fallen onto the tracks. This assumes automatic switching of the surveillance monitor or automatic connection to an emergency brake system.

At Japan Railway's Ushiku station, scenes of people falling and situations assumed dangerous are captured on Platform 3. The actions of falling from the platform, sitting, sleeping, neglect of baggage on the platform, and so forth were recorded for both the platform edge and the railroad track edge. We also recorded some boxes and chairs as examples of baggage left on the platform. It was confirmed that the existence of a person on the platform could be clearly detected at all times.

For a platform edge monitoring experiment at Tokyu Yokohama station, five stereo cameras were placed in a row at the platform edge of Track No. 1, and the coming and going of actual passengers were captured continuously from the first to the last train during the day. The captured range overlapped between adjacent cameras, and the length of one car (18 m) was completely covered. Figure 6.9 shows an entire 3D reconstruction of the platform, while Figure 6.10 shows the passenger recognition results during the morning rush hour, at 8 s.m The recognition was most

Fig. 6.9 3D reconstruction of a platform edge.

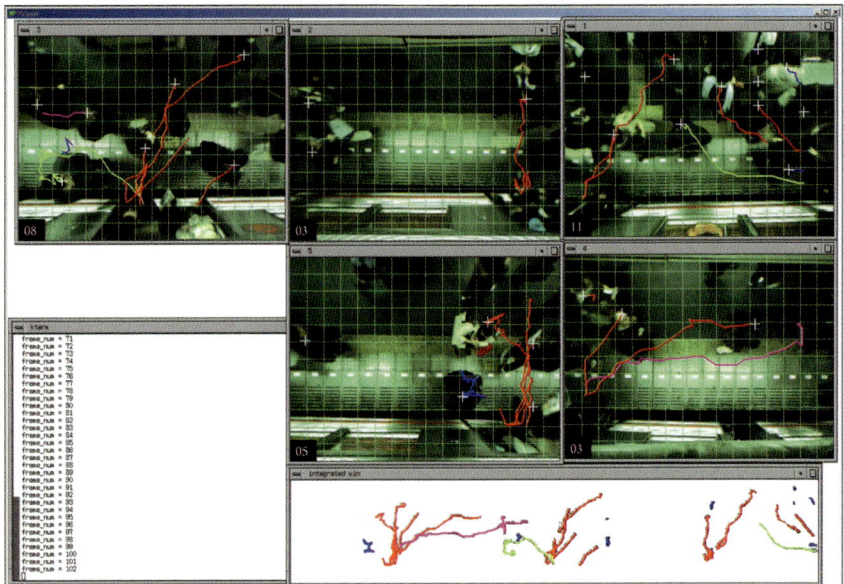

Fig. 6.10 Traffic management by using USV.

difficult when the morning sun created silhouettes. Through the recognition from the stereo range images, we confirmed that the system could distinguish crowds of passengers.

6.5.3 Monitoring Huge Space

Our monitoring technology is meant not only for safety, but is also adaptable to trajectory tracking in a huge open space. We applied our vision system at Aichi Expo 2005, in the Orange Hall of the Global House. We used four stereo camera servers to see the floor from the ceiling of the hall. The system monitored the trajectories of all attendees for a half year. The main purposes of this experiment were the following:

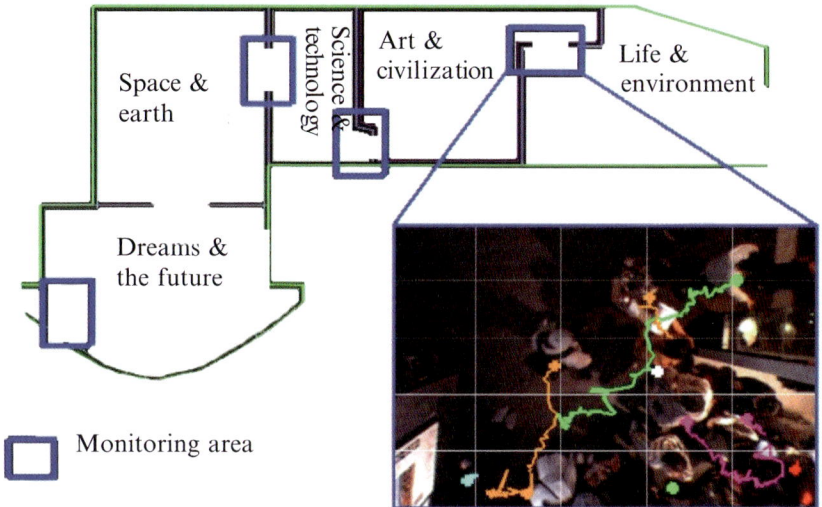

Fig. 6.11 Orange Hall of the Global House at Aichi Expo 2005.

1:      Counting the number of people in each large block in real time;
2:      Analyzing behavior through the camera; and
3:      Experimenting simultaneously with RF-ID technology (*Aimulet*).

Figure 6.11 shows a schematic map of the hall, including the monitored areas, and the tracks through one area. Because all attendees had to pass through the four monitored areas, the system could count the numbers of people inside the three large areas in real time.

Because all sensors have the advantages and disadvantages of sensing, we are also researching combinations of vision and other sensing methods. We consider this research field to offer great possibilities and great promise.

6.6 Conclusion and Future Work

We have described the basic concepts of Ubiquitous Stereo Vision and examples of real applications. The purpose of our research is to obtain information by statically placing stereo cameras in a location. In other words, we are developing a useful technique that simultaneously utilizes 3D and color image information obtained at a semi-video rate. We have explained the hierarchical utilization of three-dimensional information, a personal segmentation method, and a personal expression method as fundamental techniques for this technology.

In addition, we have shown specific cases of indoor space control, safety management, and tracking in a huge space. As another example, we aim to develop a vision interface for manipulating all objects in a specific working area. We also aim to develop applications such as safety management and traffic management for public spaces. Furthermore, we can apply this method to measuring the movements of many people in a huge space. We are thus developing an automatic learning method, an object description method, and a recognition method for actual locations involving people.

References

1. I. Yoda and K. Sakaue: "Utilization of Stereo Disparity and Optical Flow Information for the Computer Analysis of Human Interaction," Machine Vision and Applications, Vol. 13, No. 4, pp. 185–193, Mar. 2003.
2. T. Matsuyama: "Cooperative Distributed Vision," Proc. 4th International Workshop on Cooperative Distributed Vision, pp. 1–25, Mar. 2001.
3. T. Kanade, P. W. Rander, and P. J. Narayanan: "Virtualized Reality: Constructing Virtual Worlds from Real Scenes," IEEE Multimedia, Vol. 4, No. 1, pp. 34–47, 1997.
4. I. Kitahara, Y. Ohta, H. Saito, S. Akimichi, T. Ono, and T. Kanade, "Recording of Multiple Videos in Large-scale Space for Large-scale Virtualized Reality," Proceedings of International Display Workshops (AD/IDW'01), pp. 1377–1380, 2001.
5. W. Matusik, C. Buehler, R. Raskar, S. J. Gortler, and L. McMillan: "Image-Based Visual Hulls," Proceedings of the 27th Annual Conference on Computer Graphics and Interactive Techniques, pp. 369–374, 2000.
6. J. Krumm, S. Harris, B. Brumitt, M. Hale, and S. Shafer: "Multi-Camera Multi-Person Tracking for Easy Living," Proc. International Workshop on Visual Surveillance, pp. 3–10, 2000.
7. M. H. Yang, D. J. Kriegman, and N. Ahuja: "Detecting Faces in Images: A Survey," IEEE Transactions on PAMI, Vol. 24, No. 1, pp. 34–58, Jan. 2002.
8. FaceIt: http://www.identix.com/
9. A. Pentland: "Looking at People: Sensing for Ubiquitous and Wearable Computing," IEEE Transactions on PAMI, Vol. 22, No. 1, pp. 107–119, Jan. 2000.
10. Y. Yamauchi, I. Mihara, and M. Doi: "Proposal Experiments for Hand's 3D Posture Detection for Real-time Human-computer Interaction," IPSJ Journal, Vol. 42, No. 6, pp. 1290–1298, June 2001.
11. M. Yamamoto and K. Yagishita, "Scene Constraints - Aided Tracking of Human Body," Proceedings of IEEE International Conference on Computer Vision and Pattern Recognition, pp. 1151–1156, 2000.
12. I. Yoda, K. Yamamoto, D. Hosotani, and K. Sakaue: "Human Body Sensing Using Multipoint Stereo Cameras," Proceedings of International Conference on Pattern Recognition 2004 (ICPR 2004), Vol. 4, pp. 1010–1015, 2004.8.
13. Y. Yamamoto, I. Yoda, and K. Sakaue: "Arm-Pointing Gesture Interface Using Surrounded Stereo Cameras System," Proceedings of International Conference on Pattern Recognition 2004 (ICPR 2004), Vol. 4, pp. 965–970, 2004.8.
14. I. Yoda, Y. Sato, and K. Sakaue: "Automatic Face Classification and Recognition Using Self-organization," Proceedings of International Conference on Pattern Recognition 2004 (ICPR 2004), Vol. 4, pp. 1006-1009, 2004.8.

15. I. Yoda, D. Hosotani, and K. Sakaue: "Multi-point Stereo Camera System for Controlling Safety at Railroad Crossings," Proc. of the IEEE International Conference on Computer Vision Systems, 2006.1 (in print).
16. I. Yoda, D. Hosotani, and K. Sakaue: "Ubiquitous Stereo Vision for Controlling Safety on Platforms in Railroad Stations," Proc. of the Sixth Asian Conf. on Computer Vision (ACCV 2004), Vol. 2, pp. 770–775, 2004.1.

Chapter 7
Augmenting Professional Training, an Ambient Intelligence Approach

B. Zhan, D.N. Monekosso, S. Rush, P. Remagnino, and S.A. Velastin

Abstract This chapter presents interdisciplinary research work carried out at Kingston University, as a joint effort between the Faculties of Computing and Nursing. The cross-Faculty project aims at developing algorithms able to automatically interpret behavior in an extremely complex scene. The application is the professional training of student nurses and medical students, carried out in a large simulation where actors play the role of patients while instructors test individual and group medical skills of students. The chapter introduces the problem, the experimental setup and discusses some of the implemented algorithms for behavior analysis. In the context of our project, the Ambient Intelligence paradigm is interpreted as a set of guidelines to develop algorithms capable of interpreting behavior in a very complex environment monitored by an array of cameras. Intelligent algorithms were studied to enhance and automate the professional training of nurses.

7.1 Introduction

This chapter summarizes research carried out for an interdisciplinary project to aid professional skills' practitioners at Kingston University [1]. The project has engaged the computer vision team in the Faculty of Computing, Information Systems and Mathematics and the School of Nursing at Kingston University.

The School of Nursing at Kingston Hill campus trains student nurses, paramedic and medical students (in a joint degree with St. George's Medical School, London).

B. Zhan, D.N. Monekosso, P. Remagnino, and S.A. Velastin
Faculty of Computing, Information Systems and Mathematics, Kingston University,
e-mail: B.Zhan@kingston.ac.uk

S. Rush
School of Nursing, Kingston University, e-mail: srush@hscs.sgul.ac.uk

[1] The research was partially funded by the European Office of Aerospace Research and Development (EOARD) project FA8655-06-1-3013.

D. Monekosso et al. (eds.), *Intelligent Environments*, Advanced Information
and Knowledge Processing, DOI: 10.1007/978-1-84800-346-0_7,
© Springer-Verlag London Limited 2009

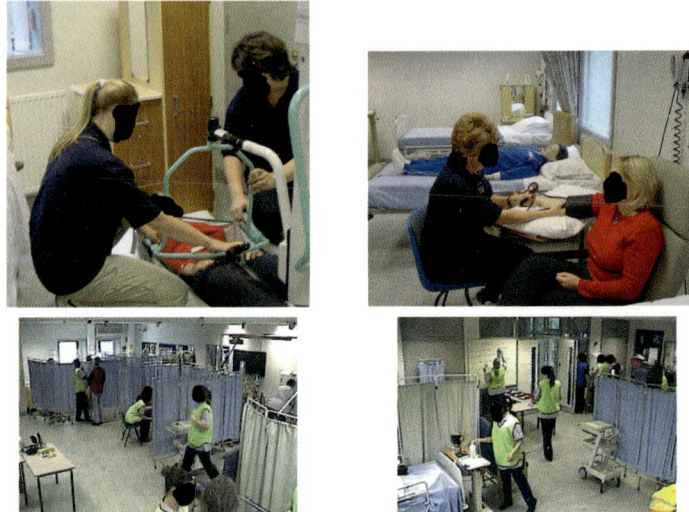

Fig. 7.1 Pictures illustrating two individual skills and two instances of a typical simulation.

The training consists of individual and group practical exercises based on taught techniques (Figure 7.1 illustrates examples of individual and team skills), entailing both medical and managerial skills. Group skills are tested in large simulations. During term time, practice skills training is organized in series of morning and afternoon sessions. Simulations involve a preliminary preparatory roundtable discussion to introduce the practical exercises, the actual simulation where skills are tested at individual and team level and a final roundtable discussion, where strengths and weaknesses of the assessed students as individuals and groups are discussed.

Intelligent algorithms were studied to enhance and automate the professional training of nurses. The inter-faculty project is the first attempt at Kingston University to design an Ambient Intelligence system, for use in the training of professionals. Ambient Intelligence is a paradigm introduced by the European community in 2000 [10], to describe a user-centric intelligent system, capable of serving the generic or specific user, responding to the needs of the individual and the group. In the context of our project, the paradigm is interpreted as a set of guidelines to develop algorithms capable of interpreting behavior in a very complex environment monitored by an array of cameras. At present the implemented algorithms do not incorporate the necessary user feedback, but this will be part of future work.

Conventional training of nurses and medical students is very time consuming and when large numbers of students are involved, it is very hard for an instructor to assess correctly the performance of a student or a group of students. The School of Nursing runs state-of-the-art training methodology, engaging students in individual and team work. Assessment is usually carried out during the practice with on-the-fly verbal feedback but also by recording video footage of students' performance,

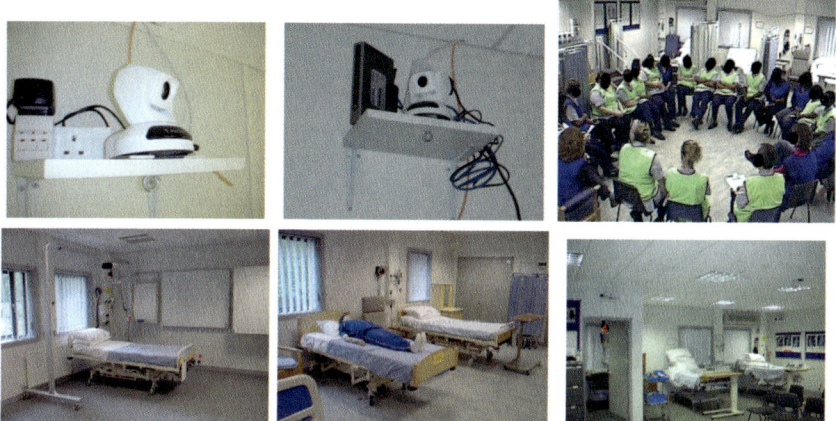

Fig. 7.2 Pictures of the experimental setup, including two pan-tilt-zoom (PTZ) cameras, the used router, some views of the skills' laboratory and an example of roundtable meeting.

discussed in classes to illustrate best practice, encouraging less capable students, praising best practice of better students. The Skills' laboratory situated at Kingston Hill campus at Kingston University can host up to 30 students at a time with instructors and role players engaged in large simulations. The lab is currently endowed with a variety of medical equipment and mobile and fixed cameras. The images in Figure 7.2 illustrate the experimental setup, the used skills' large laboratory (medically equipped), and a roundtable example.

The inter-faculty collaboration was established in 2001; it started thanks to a common research interest on human behavior in complex scenes. Both partners were driven by complementary research interests: the nursing practitioner interested in an innovative educational methodology using video recordings and the computer vision team interested in studying algorithms to describe automatically a scene, in terms of human dynamics.

Computer vision techniques used in monitoring applications lend themselves well to the automatic understanding of semantics (identification, classification and dynamics explanation of a simulation) in a professional training environment. Automatic understanding of scenes has been studied in [5] where the scene understanding is achieved through the creation of event models and in [11], where behavior profiles are built to identify anomalous behavior. In [7] and [9] semantic information is employed to cluster and index the video data. Our application bears resemblance to monitoring applications, as all scenes are extremely complex and the main goal is to model nominal behavior (best practice) and deviations thereof (bad practice). The objectives of this project, described in this chapter, include the identification and classification of role players and algorithms to describe the dynamics in the environment.

The algorithms described in this chapter are tested on video data where all role players in the scene wear a colored tabard. Four colors are used to distinguish among instructors (blue), student nurses (yellow), medical and paramedic students (green) and patients (red). The color coding was introduced to simplify the computer vision processes. In our experiments, we have employed four cameras (pan-tilt-zoom used as fixed cameras). A preliminary study was carried out by analyzing the four views independently, attempting at generating the automatic understanding of an evolving scene.

Section 7.2 describes the algorithm used to track people in the environment, Section 7.3 describes a simple algorithm employed to provide a coarse count of people in the environment and the algorithm designed to deliver an automatic reasoning about the scene. Section 7.4 illustrates some results and Section 7.5 summarizes the proposed method and introduces some future work.

7.2 Color Tracking of People

Color models have been used in computer vision research to recognize and track people and objects of interest. In particular, color models are trained by using video data of a given color using template patches, for instance, using the expectation maximization algorithm [4]. A color model is fairly robust to changes in illumination but it has the weakness of being specific to a given camera. In all our tests, each camera we used to acquire video data was color calibrated (i.e., a color model for each specific camera was built). Color calibration is an off-line process and does not affect the overall performance of the algorithms deployed for recognition and tracking. Color models were trained for the four different colors used to recognize the categories of people. These include the student nurse (yellow), the instructor (blue), the patient (red) and the medical student (green).

In order to track color patches, we have implemented the CAMSHIFT algorithm originally proposed in [1], as an evolution of the MEANSHIFT algorithm [3, 8]. CAMSHIFT adapts to the evolution of a probability density function (PDF) by alternating cycles of the MEANSHIFT algorithm with a resizing of the search window. The window size is a function of the center of mass of the probability density map (zeroth moment).

Tracking color patches entails running the CAMSHIFT algorithm for each patch. However, this is not sufficient to maintain hypotheses in a rapidly evolving scene. That is why our method keeps track of a list of alive patches, by tracking them throughout the scene with the CAMSHIFT algorithm, removing those which have too low a probability associated for a number of frames and introducing new patches, whenever sufficiently large new patches appear in the scene with a sufficiently high probability. More details of the developed algorithm can be found in [2].

7.3 Counting People by Spatial Relationship Analysis

Color segmentation generates fragmentation, by identifying one person with more blobs. Segmentation could also cause false groupings, by clustering together more people in close proximity. Both problems are due to occlusions (between people and objects) and self-occlusions (between people body parts), and also by the reflections of artificial illumination on the monitored person.

In our algorithm spatial relationships group the blobs split from a single person. At first, for each frame a graph is created with links between all identified blobs. Each link is then evaluated to judge whether the linked blobs should be merged into a cluster to recover an individual or they should be kept separate, making the assumption that both blobs are disjoint, likely to be part of different people in the scene.

First we describe a simple algorithm that can provide a qualitative counting of the people acting in the monitored scene. We will then deal with a more elaborate algorithm, whose performance is also quantified using conventional performance measures.

7.3.1 Simple People Counting Algorithm

As mentioned previously in this chapter, one of the main problems caused by the color segmentation is the fragmentation or over-segmentation of people in the scene. When a person is close to the camera, it is usually represented by a number of blobs bearing the same color.

One way of solving the problem is by grouping the blobs, using a proximity constraint. A first attempt at providing the user with a rough count of people in the scene can be done by employing an accumulator along the horizontal axis of the scene. Such accumulator will accrue information of existing blobs of a given color. The implemented algorithm simply accumulates vertically the contributions of each blob and adds such contributions to the accumulator. This is illustrated in Figure 7.3.

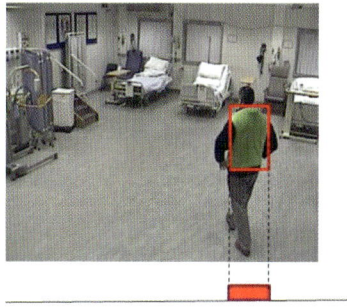

Fig. 7.3 The bounding box of a blob representing a person or part of a person is collapsed onto the horizontal axis. It will contribute to the profile of the scene for that specific category of people.

The rationale is that more blobs in close neighborhood contribute to peaks in the 1D signature, and that the likelihood of blobs belonging to people next to one another is lower than the likelihood of belonging to the same person. The algorithm simply accumulates over time the blobs identified in the video sequence and it normalizes the signature to a given maximum height. The signature is then smoothed a few times with a Gaussian filter, and the modes are identified on this signature, as the highest peaks. The signature works effectively as a probability density function of the presence of blobs in the scene. Peaks which are suboptimal, as closer to higher peaks, are eliminated, removing false alarms, and peaks which are sufficiently close are merged together by the Gaussian smoothing, effectively integrating information.

By no means this can be claimed a perfect method. In fact, it clearly suffers from the loss of vertical dimension, collapsing vertically each blob, therefore losing the information of *how far* a person is in the scene. The algorithm also underestimates the people count, by suppressing peaks that may be small, but still identify the presence of a person in the scene. The sparseness of blobs oversegmenting a person, could also introduce noise and identify more people than there are in the scene.

Figure 7.4 illustrates the pros and cons of the developed algorithm. In the following, the frames in Figure 7.4 are referred to using an incremental numbering, starting from the top left with frame 1. In frames 1, 3 and 15, people are isolated and thus the algorithm is successful. In frame 11 for instance, the color segmentation fails and introduces false alarms, which are in turn identified as peaks in the related PDFs. Frames in which people are at different distance from the camera, but not aligned can be correctly interpreted as shown in frames 13 and 14. In other cases, the algorithm fails to disambiguate perfectly aligned people as shown in frames 2 and 9. The algorithm might fail to detect people in the scene, due to illumination problems or because people are too far from the camera, as shown in frame 1.

7.3.2 Graphs of Blobs

Graphs are generated from the previously detected blobs. The nodes in the graph represent blobs while the links in the graph joining pairs of blobs represent the spatial relationship between the two blobs. The creation, deletion and updating of the links are required to be automatic according to the change of the situation. The algorithm we developed enforces links between a blob, say A, with all the other blobs in the scene during its life cycle. During the life cycle of A, another blob, say B, could appear in the scene and then leave the scene. Under such circumstance, A should then be linked to B once B has entered and the link should be eliminated right after B has left the scene. The complexity of the problem increases when the number of people involved increases. The creation of the links are triggered by the appearance of blobs, deletions are triggered by the disappearance of blobs, while

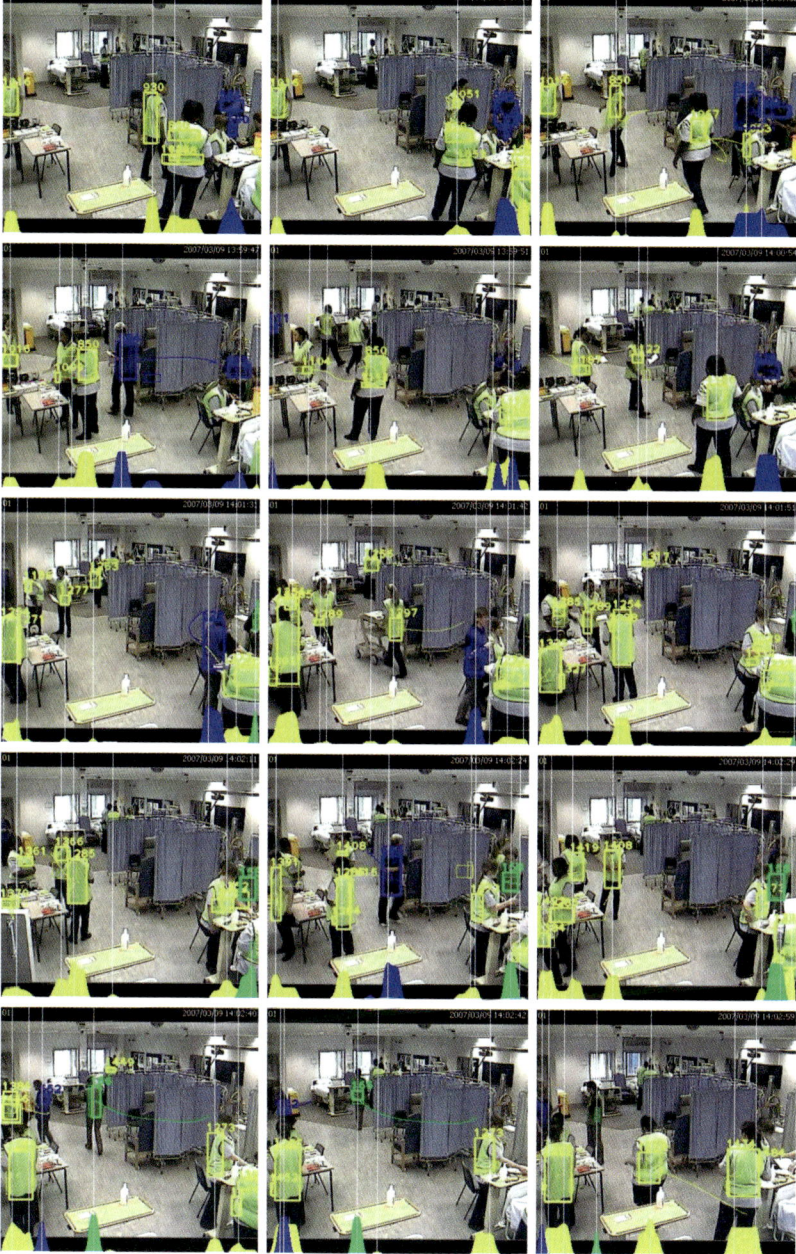

Fig. 7.4 From top left to bottom right frames are numbered frame 1 to frame 15. The above figure illustrates fifteen frames. The frames include the bounding rectangles, detected by our color tracker, and the profiles representing the probability density functions of the defined categories of role players. The white vertical lines illustrate the detected peaks, corresponding to an estimate of the modes. Each mode represents a person in the monitored scene.

updating is carried out at regular intervals every Δt, taking into consideration all the blobs at that moment in time.

Following with the above example, a link is created between A and B, when B enters the scene. The link should be kept updated while B is in the scene. The link should then be removed when B is no longer in the scene. For algorithmic simplicity, a link is bidirectional, so each link between blob A, for instance, and any other blob, also implies that all linked blobs keep track of the existence of A. When a blob leaves the scene, it sends a signal to all the links connected to it, to release and delete them. At each frame, sampled at a given Δt, the system checks the blobs to create, delete or update the existing links. Algorithm 7.1 illustrates this process.

Algorithm 7.1 The creation, deletion and update of the links

---

if objects: $O_{0...m}^{-}$ are leaving the scene **then**

    for $i = 0$ to m **do**
        Object O_i^{-} send signals to all the links connected with it
        Delete O_i^{-}
    end for
end if
Delete links with signals
if objects: $O_{0...n}^{+}$ are entering the scene **then**

    for $j = 0$ to n **do**
        Build links between object O_j^{+} and all of the existing objects in the scene
    end for
end if
Update all of the existing links

7.3.3 Estimation of Distance Between Blobs

Spatial relationships between blobs are represented as distance information contained within the links connecting nodes. The distance between blobs is calculated as the Euclidean distance between the blobs' centers. Because of the perspective distortion, the absolute value of the Euclidean distance cannot be used to estimate the spatial relation between the blobs. For instance, two blobs at an absolute distance of 50 pixels could be close to each other when they are in front of the camera while they could be far from each other when they are distant from the camera. Hence, a method for calculating relative distance by comparing the absolute distance with the size of connected blob has been proposed here, i.e., the ratio of the absolute distance and the blob size was used. In this method the variation of dimensions of blobs at different locations is considered. The Euclidean distance used as the absolute distance between blob i and j is as below:

$$D_{ij} = \sqrt{(x_i - x_j)^2 + (y_i - y_j)^2} \qquad (7.1)$$

where (x_i, y_i) and (x_j, y_j) are the coordinates of center points of blob i, j, respectively, and the temporal relative distance of blob i and j is calculated as

$$d_{ij} = \frac{D_{ij}}{\sqrt{w_k^2 + h_k^2}}, \quad k = \begin{cases} j, & \text{if } y_i - 0.5h_i < y_j - 0.5h_j \\ i, & \text{otherwise.} \end{cases} \qquad (7.2)$$

where w_k and h_k are the width and height of the blob. The denominator is a measure of the size (its diagonal) of the blob and is used as a weight, as a compensating factor for the link.

The above calculations are carried out in a single frame. A temporal average operator has been applied every Δt frames for each distance calculation. This operation can reduce the instability caused by the tracking algorithm, thus the video sequence has been sampled at fixed regular time intervals, i.e., each time segment contains distance information for Δt frames. Equation (7.3) describes the calculation of this distance,

$$\bar{d}_{ij}(T) = \frac{1}{\Delta t} \sum_{\Delta t} d_{ij}(T - \Delta t) \qquad (7.3)$$

so the distance between blobs i and j at time T is the average of the distances over the previous Δt frames. The main reason for this temporal smoothing operation is to stabilize the distance. Δt is a short time interval, for example in our case we use an 8-frame Δt, which is equivalent to 0.5 second.

7.3.4 Temporal Pyramid for Distance Estimation

Short-term spatial relations are not sufficient for clustering blobs. A temporal pyramid of distance scheme has been introduced to maintain longer term distance information. In our algorithm, two blobs belong to the same cluster if they are close to each other, during their life span. A coarse pyramid was used, where the current time frame is represented by the top of the pyramid, while the whole lifetime of the blob and half of its lifetime represent the other two layers. For each pair of blobs, the algorithm takes into account the distance information from each level of the pyramid and calculates the overall probability that they belong to the same cluster. This scheme is based on an assumption that two persons are not likely to stay next to one another for a very long time period. This is clearly not true in general, but it suits well the application of nurse training, where nurses, instructors and medical students are continuously moving about.

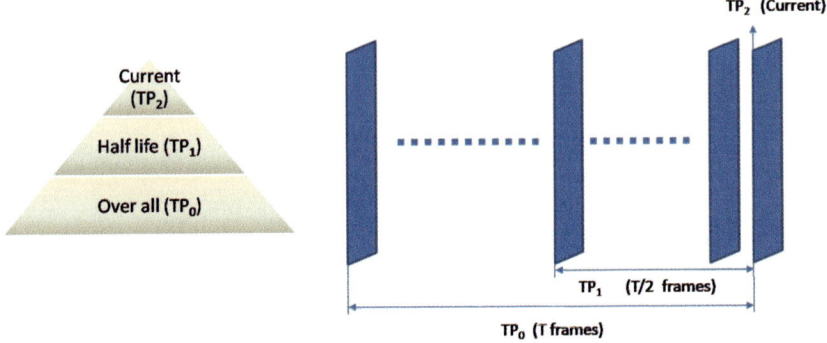

Fig. 7.5 Temporal Distance Pyramid: The bottom layer represents the overall distance information from time 0 to time T, the middle layer represents the distance information from time $\frac{T}{2}$ to T and the top layer holds the distance information for the current time slice T.

The temporal pyramid consists of three levels: the bottom layer holds the overall distance information between two blobs from their appearance in the scene to the present time, the top layer holds the present distance information and the middle layer holds the information from the half time to the present; this is illustrated in Figure 7.5. The generation of the temporal distance pyramid is:

$$TP_0(T) = \bar{d}(0 \rightarrow T) = \frac{1}{T} \sum_{t=1}^{T} \bar{d}(t) \tag{7.4}$$

$$TP_1(T) = \bar{d}(T/2 \rightarrow T) = \frac{1}{\frac{T}{2}} \sum_{t=\frac{T}{2}}^{T} \bar{d}(t) \tag{7.5}$$

$$TP_2(T) = \bar{d}(T) = \bar{d}(T) \tag{7.6}$$

where $TP_0(T)$ to $TP_2(T)$ represents the distance information held from the bottom layer to the top layer at time T. In practice to reduce the redundant calculations of top layer ($TP_0(T)$) and middle layer ($TP_1(T)$), a recursive method has been employed and the equations are modified as follows:

$$TP_0(T) = \frac{1}{T}(TP_0(T-1) \times (T-1) + \bar{d}(T)) \tag{7.7}$$

$$TP_1(T) = \frac{1}{\frac{T}{2}}(TP_1(T-1) \times \frac{T-1}{2} - \bar{d}(\frac{T}{2} - 1) + \bar{d}(T)) \tag{7.8}$$

$$TP_2(T) = \bar{d}(T) \tag{7.9}$$

7.3.5 *Probabilistic Estimation of Groupings*

A probabilistic clustering scheme was devised to eliminate over-segmentation in the scene. As mentioned earlier in the chapter, one person may be identified with more than one blob.

Clustering is carried out for each category, so, if two blobs belong to colors/categories that refer to two different role players, for instance instructor (blue) and student nurse (yellow), then their link has probability zero and they cannot be linked to the same graph. In all other cases, spatial relation is the main criterion used for clustering. This means that the probability associated with the link between blobs is inversely proportional to their Euclidean distance. This rule is represented by a function $\varphi(\overline{d})$:

$$P(\overline{d}) = \varphi(\overline{d}) = \begin{cases} 1, & when \ \overline{d} = 0 \\ 1 - \frac{1}{\theta_d} \times \overline{d}, & when \ 0 \le \overline{d} \le \theta_d \\ 0, & when \ \overline{d} > \theta_d \end{cases} \qquad (7.10)$$

where θ_d is the threshold of distance. When the distance falls below this value, the probability of clustering is equal to 0. When the distance is equal to 0, the probability is equal to 1. The probability of clustering two blobs with a distance that falls between 0 and θ_d is interpolated with a linear function. Each layer of the temporal distance pyramid provides a probability of clustering and the outcome of the three layers has been averaged as follows:

$$P_{dis} = \frac{1}{3}(P(TP_0) + P(TP_1) + P(TP_2)) \qquad (7.11)$$

The overall size of the blobs is also used to bias the probability of clustering blobs. A linear approximation of the blob size at different locations of the scene has been used as reference. The size of the overall bounding box between blobs is compared against the estimated reference, according to their locations. This comparison is represented by the ratio

$$\overline{s} = \frac{S_o}{S_r} \qquad (7.12)$$

where S_o is the size of the blobs and S_r is the reference size from the linear approximation. The probability of clustering by area is calculated by $\varphi(\overline{s})$:

$$P_{size} = P(\overline{s}) = \varphi(\overline{s}) = \begin{cases} 1, & when \ \overline{s} = 0 \\ 1 - \frac{1}{\theta_s} \times \overline{s}, & when \ 0 \le \overline{s} \le \theta_s \\ 0, & when \ \overline{s} > \theta_s \end{cases} \qquad (7.13)$$

where θ_s is the threshold of the ratio of the size ($\overline{s}$). $\varphi(\overline{s})$ is employed for the reason that smaller fragments should increase the probability to cluster. The overall probability of clustering is

$$P = P_{dis} \times P_{size} = P(\overline{d}) \times P(\overline{s}) \qquad (7.14)$$

7.3.6 Grouping Blobs

For each frame, the clustering takes place in two steps, we call *pair clustering* and *sub-clustering*. Pair clustering checks all pairs of blobs, clustering together all the pairs with high probability. This rule ensures that all the blobs which potentially belong to the same person are clustered together. If two blobs are selected to be clustered and they already belong to two clusters, then the clusters can be merged as shown in Figure 7.6(a). Pair clustering may generate bad clustering. In fact, blobs which belong to different persons could be clustered together as shown in Figure 7.6(b). The second step, sub-clustering is used to get the scores of different number n ($1 \leq n \leq N$) of sub-clusters of a cluster C which contains N blobs. In a cluster generated in the pair clustering step, each pair of blobs is associated with a probability of clustering which is generated by the method described in Section 7.3.5. The strength Γ of a cluster is defined as

$$\Gamma = \frac{1}{C_N^2} \sum_{i=0}^{C_N^2} P_i \qquad (7.15)$$

where N is the total number of blobs, so there are C_N^2 pairs of blobs. We define *Connected* and *Unconnected* as the pairs of blobs with a probability of clustering respectively higher and lower than a given threshold. Creating sub-clusters requires that, every time the weakest *Connected* link is removed, the blobs are reclustered by the remaining *Connected* list. The score of the operation is equal to the energy cost E of removing the *Connected* list and the related *Unconnected* list.

$$\Lambda = \frac{1}{n} \sum E + \frac{1}{m} \sum \Gamma \qquad (7.16)$$

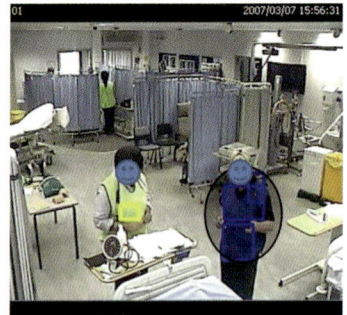

(a) A frame in which multiple blobs (illustrated with a black oval) should be clustered together.

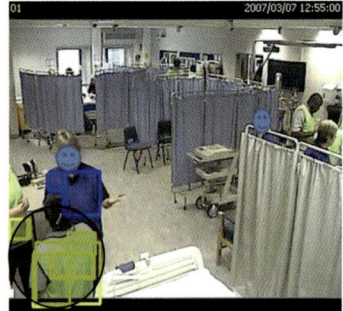

(b) A frame in which blobs belong to different persons could be clustered together (illustrated with a black oval).

Fig. 7.6 Two frames of problems in clustering.

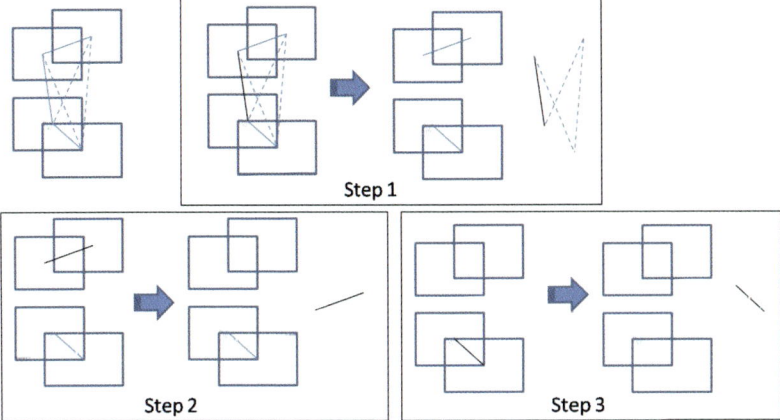

Fig. 7.7 An example of sub-clustering. Solid lines between blobs show the *Connected* and the dashed lines are the *Unconnected*. In each step, the black *Connected* is removed, and the related *Unconnected* are removed. This operation is updated until all the *Connected* are moved and all the blobs are isolated.

where the energy cost of removing a *Connected* list with probability of clustering P is

$$E = 1 - P \qquad (7.17)$$

This operation is kept to carry on until all the *Connected* are removed, meanwhile all the blobs are isolated. Figure 7.7 shows an example of the sub-clustering process of a cluster containing four blobs.

During the operation, the scores are accumulated for different number of sub-clusters. In this case the number of sub clusters with highest score is selected to be added to the person-count and the sub-clusters are regarded as individuals. The total number of people is the sum of the selected numbers of sub-clusters of all the clusters in the frame.

7.4 Experimental Results

We have tested the implemented algorithm with a number of video sequences consisting of at least 300 frames. The sequences are a selected sample from a large database of video data acquired at Kingston Hill, during a number of simulation sessions. Selected excerpts of video sequences were ground truthed.

For a video sequence, the number of people as well as their locations are retrieved for each frame. To access the system performance, ground truth is manually marked up by the ViPER Ground Truth Authoring Tool(ViPER-GT tool), which is a part of The Video Performance Evaluation Resource (ViPER) developed by the Language

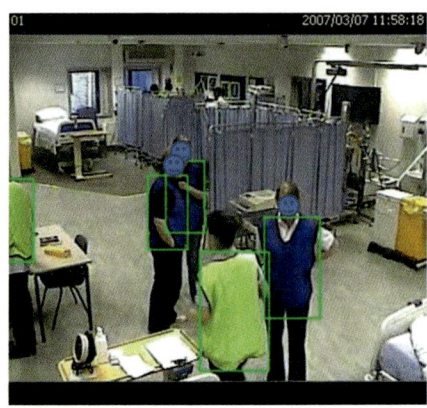

Fig. 7.8 A ground truth example from ViPER-GT.

and Media Processing Laboratory, University of Maryland[2]. The *ground truthing* process is carried out every frame, and each person is selected by a bounding box (Figure 7.8). In our work, performance was evaluated using measures borrowed from the information retrieval literature. Recall and Precision, which have been used in evaluating search strategies, are used here to test the results of our algorithm against ground truth information. Recall is the ratio between the number of relevant records retrieved and the total number of relevant records in the database. Precision is the ratio between the number of relevant records retrieved and the total number of irrelevant and relevant records retrieved. The Precision-Recall curve is employed to provide a quantitative assessment of the performance of the algorithm [6]. The bounding boxes of the ground truth *GT* and the bounding boxes generated by the presented algorithm *RE* are used to estimate the following measures:

$$Recall = \frac{GT \cap RE}{GT} \tag{7.18}$$

$$Precision = \frac{RE \cap GT}{RE} \tag{7.19}$$

Category information is also considered, i.e., intersections of *GT*s and *RE*s with different colors are not taken into account. The Recall and Precision estimates have been recorded along with time scale in all the video sequences, and each pair of the measures contributes as a point on the Precision-Recall curve. Figure 7.9 shows the Precision-Recall curve for a video sequence. The graph in Figure 7.10 illustrates the counting results from different situations with different number of people and different number of professions. These results show that the system has a stable performance under different circumstances.

---

[2] The details of ViPER and ViPER Ground Truth Authoring Tool are available online at http://viper-toolkit.sourceforge.net/.

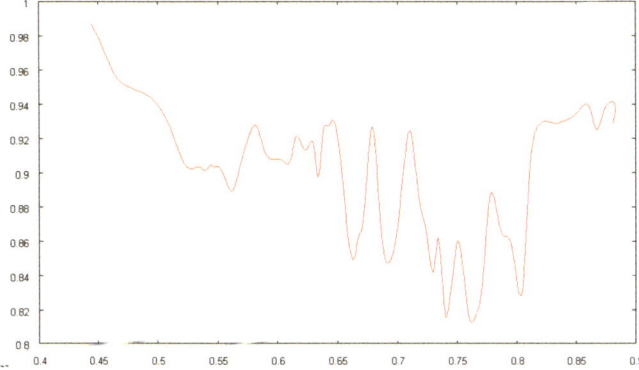

Fig. 7.9 Precision-Recall curve.

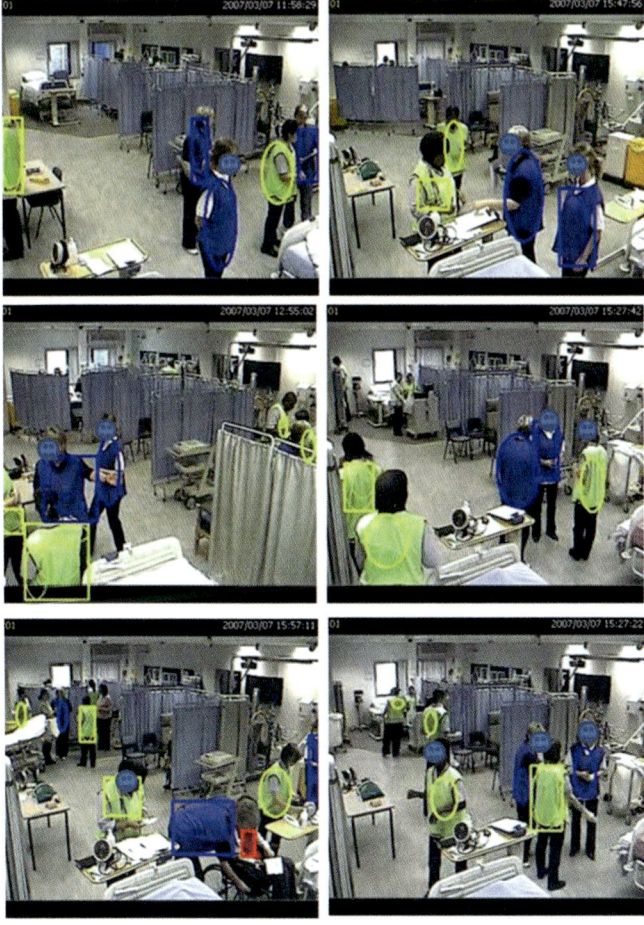

Fig. 7.10 Counting people: In these figures, ellipses represent the original blobs; thick outlines of shapes (rectangles, ellipses) show the existences of individuals; thin outlines of ellipses show the existences of clustered blobs.

7.5 Conclusions

This chapter has described an intelligent system that follows the guidelines of the Intelligent Environment paradigm. At present, only cameras are used to recognize behavior and estimate the category and number of people in the scene. Color models are used to track people in the scene and provide sufficient information for the system to generate graphs of detected and tracked color patches. The generated graphs are then automatically analyzed by an algorithm, to cluster blobs and estimate the number of people in the scene.

The major challenge is to identify individuals from the color segments in a complex dynamics environment. The basic assumption is that two individuals will not always remain in close proximity over a long period of time. The current system provides a fairly good estimation of number of people. High precision values in quantitative results suggest that the system has a low false alarm rates. This is also confirmed by observing the qualitative results. However as the analysis is limited to 2D information, the system would fail to count a person when seriously occluded and most of their patch is not visible from the view. As a result recall sometime drops to relatively low values. To tackle the problem of miss-counting, the next step for performance improvement is either to introduce an occlusion handling scheme or to fuse information from different views.

The contribution of this chapter is in the design of a robust algorithm for the interpretation of a complex scene. In terms of algorithm development future work will fpcus on the description of level of cluttering of the scene and dynamics descriptions of the scene such as descriptions of people interactions. Also, evidence from all the cameras can be combined to provide 3D information. In terms of technology, radio frequency technology will be introduced to help with the recognition of positional information of scene actors.

References

1. Bradski, G.: Computer vision face tracking for use in a perceptual user interface. Intel Technology Journal **2**, 1998
2. B. Zhan, N.D. Monekosso, T. Rukhsana, P. Remagnino, Y. Kuno, A. Mansur: Skin patches trajectories as scene dynamics descriptors. In: International Association of Pattern Recognition Conference on Machine Vision Applications 2007, pp. 315–318
3. Cheng, Y.: Mean shift, mode seeking, and clustering. IEEE Transactions on Pattern Analysis and Machine Intelligence **17**(8), 790–799 (1995)
4. C.M. Bishop: Pattern Recognition and Machine Learning. Springer (2006)
5. Cupillard, F., Bremond, F., Thonnat, M.,: Group behavior recognition with multiple cameras. In: Proceedings of the Workshop on Applications of Computer Vision, (ACV 2002), pp. 177–183 (2002)
6. Davis, J., Goadrich, M.: The relationship between precision-recall and roc curves. In: Proceedings of the 23rd International Conference on Machine learning, pp. 233–240. ACM (2006)
7. Fan, J., Luo, H., Elmagarmid, A.: Concept-oriented indexing of video databases: toward semantic sensitive retrieval and browsing. IEEE Transactions on Image Processing **13**(7), 974–992 (2004)

8. Fukunaga, K., Hostetler, L.: The estimation of the gradient of a density function, with applications in pattern recognition. IEEE Transactions on Information Theory **21**, 32–40 (1975)
9. Hu, W., Xie, D., Fu, Z., Zeng, W., Maybank, S.: Semantic-Based Surveillance Video Retrieval. IEEE Transactions on Image Processing, **16**(4), 1168–1181 (2007)
10. P. Remagnino, G.L. Foresti, T. Ellis (eds.): Ambient Intelligence: a Novel Paradigm. Springer (2004)
11. Xiang, T., Gong, S.: Video Behavior Profiling for Anomaly Detection. IEEE Transactions on Pattern Analysis and Machine Intelligence, **30**, 893–908 (2007)

Chapter 8
Stereo Omnidirectional System (SOS) and Its Applications

Yutaka Satoh and Katsuhiko Sakaue

Abstract The Stereo Omnidirectional System (SOS) is a completely new camera system that consists of 36 cameras arranged in the shape of a ball, only 11.6 cm in diameter. It can be used to acquire color images in all directions and obtain range data in real time with no dead angles. Since the SOS has a spherical field of view with no dead angles, it can recover completely from any camera rotation. And because its field of view is unrestricted, the SOS does not suffer from the problem when an object being observed leaves the image frame. Therefore, the SOS is perfect for applications in robot vision systems and security systems. This paper discusses the basic technology underlying the SOS and gives examples of applications.

8.1 Introduction

A Stereo Omnidirectional System (SOS) is a novel camera system capable of capturing omni-directional color images and range data simultaneously and in real time with a complete spherical field of view.

The greatest feature of the SOS is that its field of view is unrestricted, unlike those of conventional camera systems. Ever since cameras were invented, a restricted field of view has been accepted as inevitable. However, it is clearly better if there is no such restriction due to structural limits. The SOS represents a new concept in that it is not limited by a restricted field of view.

Several designs have previously been proposed to provide similar functionality with hyperbolical mirror omni-directional camera systems [1, 2, 3, 4, 5, 6]. The

Yutaka Satoh

National Institute of Advanced Industrial Science and Technology, Tsukuba, Japan,
e-mail: yu.satou@aist.go.jp

Katsuhiko Sakaue

National Institute of Advanced Industrial Science and Technology, Tsukuba, Japan,
e-mail: k.sakaue@aist.go.jp

D. Monekosso et al. (eds.), *Intelligent Environments*, Advanced Information and Knowledge Processing, DOI: 10.1007/978-1-84800-346-0_8,

SOS offers the following advantages over such systems: (1) spatial information is captured evenly and at high resolution, due to the large number of cameras: (2) complete absence of blind spots in any direction, due to a spherical construction; and (3) simultaneous, real-time capture of color images and range data. These features of the SOS enable the construction of applied systems of higher definition than previous designs.

The chapter proceeds as follows: Section 8.2 presents an outline of the SOS. Section 8.3 suggests a way of integrating individual images. Section 8.4 considers rotation recovery of omni-directional images. Section 8.5 discusses some applications of the SOS, and Section 8.6 concludes the chapter.

8.2 System Configuration

Figure 8.1 shows the external appearance of the SOS. As a completely novel camera system developed by the authors [7, 8], the SOS realizes an extremely compact design (with a diameter of 11.6 cm and a weight of 600 g) while having the capability to completely capture color images from the entire surroundings and obtain range data in real-time. The basic structure of the SOS consists of a regular dodecahedron (12 faces), with a trinocular stereo camera unit located on each face (making a total of 36 individual cameras).

To ensure the accuracy of the range data, the distance between the cameras in each stereo camera unit (i.e., the stereo baseline) must be assured. However, it is a problem that the size of a camera-head becomes excessively large when stereo camera units are arranged in such a regular-dodecahedron fashion. To address this

Fig. 8.1 Stereo Omnidirectional System (SOS).

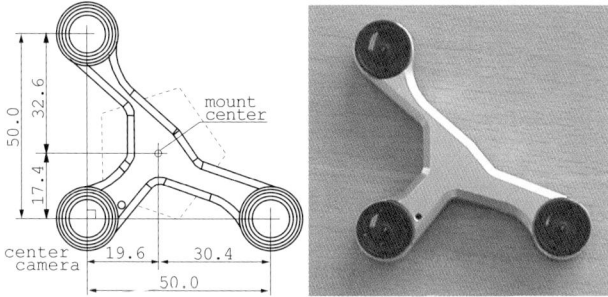

Fig. 8.2 Stereo camera unit.

Fig. 8.3 Examples of images captured by the stereo camera unit.

problem, the three cameras of each stereo camera unit (see Figures 8.2 and 8.3) are mounted on a T-shaped arm, and by arranging the base planes of the stereo camera units so that they crisscross one another, we have ensured that and downsizing is possible while keeping a constant stereo baseline.

Each camera mounted to the stereo camera unit is on the same plane, and the optical axis of each is mutually parallel to the others. And the center camera is placed at right angles to the other two cameras so that their 50-mm baselines intersect at the center camera. In this way, each stereo camera unit satisfies the epipolar constraint; as a result, the processing cost of searching for corresponding points is decreased.

Stereo calibration is performed for each stereo camera unit in advance. And for the camera heads together, we place the SOS inside a 1 cubic meter calibration box (see Figure 8.4). From inside the box the SOS observes a pattern printed on

Fig. 8.4 A calibration box. (The SOS is placed on the bottom of the box. And the bottom is covered by the left-hand side body, that is to say, the SOS is placed inside the box.)

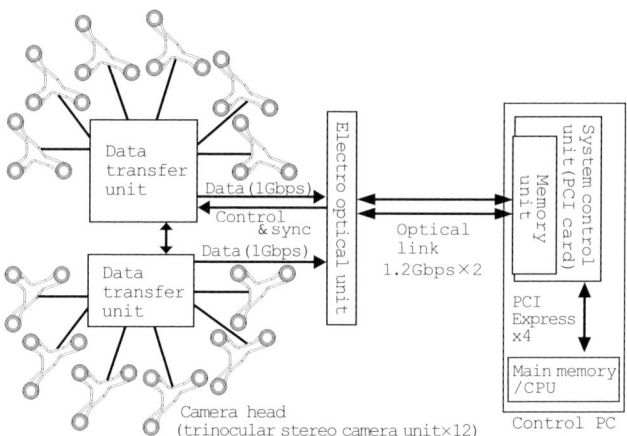

Fig. 8.5 Block diagram of the SOS.

the inside wall of the calibration box, and misalignments between camera units are eliminated by software.

The system configuration of the SOS is shown in Figure 8.5. Groups of images taken by the camera head are output as two 1.2-Gbps optical signals by means of an electro-optical conversion unit fitted in the main body. A memory unit and a control unit are mounted on one PCI-Express (x4) board, and all image acquisitions and control processes can be done by just a single PC.

8.3 Image integration

The SOS provides 12 individual color images (each image is 640×480 pixels) at 15 frames per second. Since the optical centers of constituent cameras of the SOS do not coincide, central projection image without any error makes use of range data.

For aiming at viewing, however, we can assume that the targets for observation exist in a fixed distance because compact camera head size has a small influence on inter-camera parallaxes. This assumption provides high-speed image integration as we mention in what follows.

As a prerequisite, the system calibration of the SOS must be completed and the global coordinate system with origin at the center of the SOS is defined (the internal parameters and external parameters of the cameras are known). Moreover, we assume that lens distortion compensation (rectification) for each camera is also completed.

Since the SOS covers its spherical field of view with 12 individual cameras, the image integration can be considered to be the division problem of the spherical surface.

Let S be the sphere of radius r in a coordinate system with origin at the center of the SOS (see Figure 8.6). Each vector p on S defines the camera set

$$C_p \subseteq \{1, 2, \cdots, 12\}, \tag{8.1}$$

of which each member camera has p in its FOV. Once (non-empty) C_p is given, the optimal camera c to observe p is decided by the condition

$$p \bullet n_c = max\{p \bullet n_{c'} | c' \in C_p\}. \tag{8.2}$$

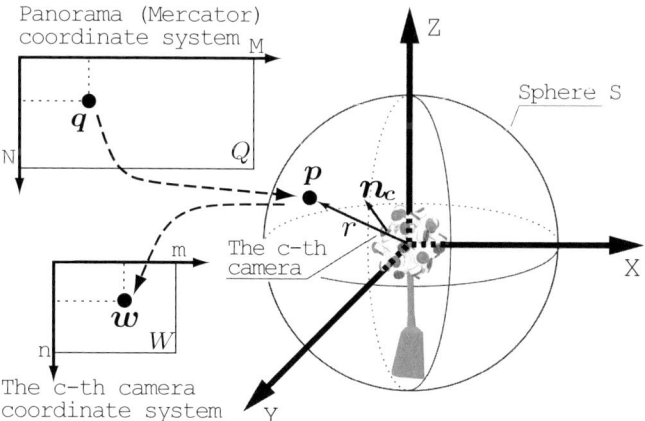

Fig. 8.6 Projection from Mercator coordinate system to a local camera coordinate system.

Note that this condition is designed to choose the best camera to observe p near the center of FOV.

Now we consider how to combine 12 images into a Mercator projection image. This argument also shows how to construct C_p. Let Q be the Mercator projection image of size $M \times N$ and let $q = q(i, j)$ be any point on Q, which is the projection of p on S. Suppose p has three dimensional homogeneous coordinate $(x, y, z, 1)^T$. Then, the following relation holds:

$$\begin{pmatrix} x \\ y \\ z \\ 1 \end{pmatrix} = \begin{pmatrix} -\cos(b)\cos(a) \cdot r \\ \cos(b)\cos(a) \cdot r \\ -\sin(b) \cdot r \\ 1 \end{pmatrix}, \tag{8.3}$$

where $a = \dfrac{2j\pi}{M}$ and $b = \dfrac{(2i - N)\pi}{2N}$.

We want to know the location (say, w) of p in the camera-coordinate system for a camera c, assuming p is included in FOV of c. The location $w = (u, v, 0, 1)^T$ in the camera image W_c can be calculated as follows:

$$w = A_c[R_c t_c]p. \tag{8.4}$$

Here, A_c is the matrix of internal parameter and $[R_c t_c]$ is the matrix of external parameter, both for the camera c. The size of the image of camera c determines whether w is actually in W_c or not. If not, we conclude $w \notin W_c$, which means $c \notin C_p$. In other words, the camera c is not the appropriate camera to observe the point p. Starting from $\{1, \cdots, 12\}$, the iterated removal of inappropriate camera gives us C_p.

Of course, if we had had the ideal SOS, we wouldn't have needed to mention C_p, and could choose the optimal camera c for p by the condition

$$p \bullet n_c = max\{p \bullet n_{c'} | 1 \leq c' \leq 12\}. \tag{8.5}$$

However, the image centers of constituent cameras of the SOS practically may have shifted from the physical design center. In such a case, the selected camera may be unable to cover point p. The above-mentioned method enables us to choose another camera to cover point p, since the SOS has overlapping regions near the boundaries between constituent cameras.

By the method explained above, we can construct the mapping $F : Q \to W_1 \cup \cdots \cup W_{12}$, which associates each q in Q to the point w in the optimal camera. Figure 8.7 shows the Mercator projection image which shows the sphere division. The numbered regions in the figure show that each of the region c belongs to the camera c. Figure 8.8 shows an example of the Mercator projection image of a real scene.

The mapping F depends only on the design of the SOS and parameters of the constituent cameras. Therefore, the mapping F can be implemented as a lookup table, which enables high-speed rendering of the Mercator projection image.

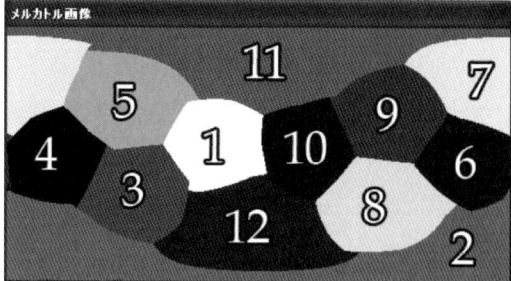

Fig. 8.7 Mercator projection image which shows the sphere division.

Fig. 8.8 An example of the Mercator projection image of a real scene.

8.4 Generation of Stable Images at Arbitrary Rotation

Since the SOS presents a complete surround view (i.e., a spherical field of view), theoretically speaking the shooting condition in response to arbitrary camera rotation stays constant. This means that if the pose of the SOS is known, it is possible to generate rotation invariant images.

To detect the pose of the SOS, motion sensors are attached to the camera support pole as shown in Figure 8.9. When the SOS is in a stationary state, the direction of the gravitational force (i.e., vertically down) is determined by means of the acceleration sensor, and the pose of the SOS is obtained. Figure 8.10 shows an example of an image whose rotation has been corrected by using the obtained pose parameters.

The upper image shows the original image (take the direction of the camera support pole as the Z axis); the lower one is the image after correction (take the vertical direction obtained by the sensor as the Z axis). It is clear that after the correction, the effect of the camera rotation has been canceled.

In the following, we investigate the arbitrary rotation of the SOS that occurs when the SOS is in motion. Under such conditions, to continually generate stable camera images, it is necessary to estimate the pose of the SOS in real time. The authors have already developed a high-speed, high-accuracy method for the pose estimation, which simultaneously utilizes sensors and omni-directional images [9]. Due

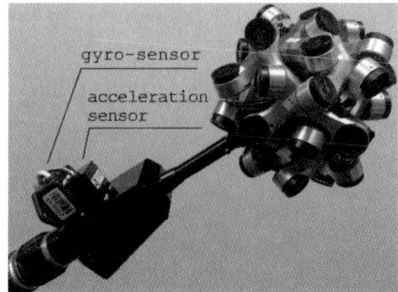

Fig. 8.9 Motion sensors.

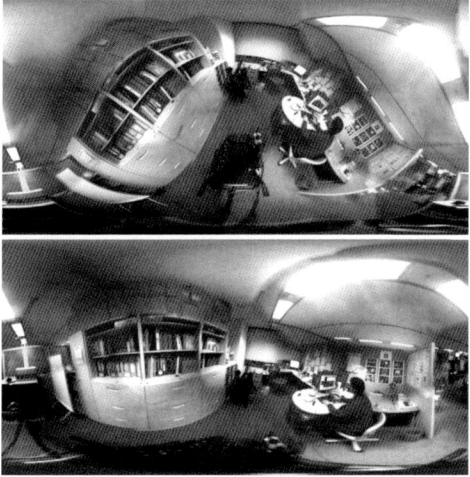

Fig. 8.10 Rotation recovery. The original image (top) and the corrected image (bottom).

to limitations of space, only the method for generating rotation-corrected images in real-time is described here.

In the case that the pose is not changing, integration of individual images into omni-directional panoramic images (as described in Section 8.3) is enabled at high speed by utilizing a pre-prepared lookup table based on fixed-orientation parameters. In the case that the pose is changing, however, the orientation parameters are unknown beforehand, so such a pre-prepared lookup table cannot be prepared and geometric-conversion processing, which incurs an extremely high cost, must be performed on each frame. To address this processing issue, we use the method described next. By means of this method, it is possible to perform correction of arbitrary camera rotation simply by using C++ pointer operation and referring to lookup tables three times.

A general outline of this procedure is shown in Figure 8.11. Here we represent the rotation of the SOS with the rotation angles of α, β and γ around the axes of X, Y and Z, respectively. If the omni-directional spherical image when the rotation changes is

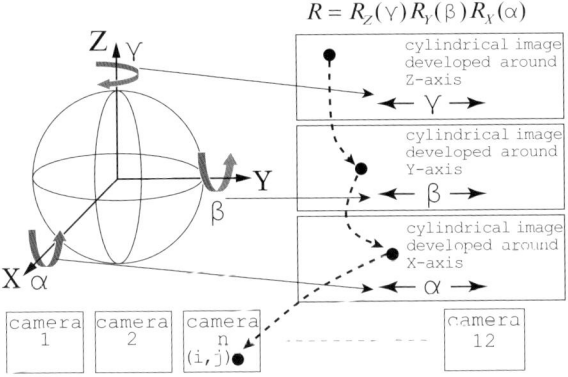

Fig. 8.11 Generation of rotation-corrected images in real time.

expressed in the three-axis angular coordinate system (A, B, Γ), the rotation invariant spherical image can be recovered as $(A - \alpha,\ B - \beta,\ \Gamma - \gamma)$. Utilizing this principle, we can generate a rotation invariant cylindrical expansion of the spherical image efficiently.

First, the correspondence relations between the cylindrical image (θ_x, ϕ_x) expanded around X-axis of the global coordinate system of the SOS and the images of each camera of the SOS. These correspondence relations are created by planar projection between camera coordinate system and the global coordinate system of the SOS, and are represented as $c_x(\theta_x.\phi_x), i_x(\theta_x.\phi_x)$ and $j_x(\theta_x.\phi_x)$. Here c_x is the camera number and i_x and j_x are the coordinates of the image of camera c_x. Next, the correspondence relations between the cylindrical images (θ_y, ϕ_y) and (θ_x, ϕ_x) expanded around Y-axis and X-axis, respectively, are computed, and are represented as $x_\theta(\theta_y, \phi_y)$ and $x_\phi(\theta_y, \phi_y)$. Similarly, the correspondence relations between the cylindrical images (θ_z, ϕ_z) and (θ_y, ϕ_y) expanded around Z-axis and Y-axis, respectively, can be obtained as $y_\theta(\theta_z, \phi_z)$ and $y_\phi(\theta_z, \phi_z)$. Using those correspondence relations, we can recover a rotation invariant cylindrical expansion of the spherical image with full FOV from the rotation parameters α, β and γ of the SOS, by adopting a multiplex indexing strategy. The following formulas show the correspondence relations of camera numbers:

$$c_y(\theta_y, \phi_y) = c_x(x_\theta(\theta_y, \phi_y) - \alpha, x_\phi(\theta_y, \phi_y)) \tag{8.6}$$

$$c_z(\theta_z, \phi_z) = c_y(y_\theta(\theta_z, \phi_z) - \beta, y_\phi(\theta_z, \phi_z)) \tag{8.7}$$

$$c_r(\theta, \phi) = c_z(\theta - \gamma, \phi) \tag{8.8}$$

In the same way as for the above formulas, the multiplex index tables can be built for $i_r(\theta, \phi)$ and $j_r(\theta, \phi)$ as well.

According to the procedure above, a PC fitted with dual 3.6-GHz CPUs can generate images corrected for rotation in about 15 ms.

8.5 An example Application: Intelligent Electric Wheelchair

8.5.1 Overview

Since the capability of the SOS is extremely powerful, it is expected to have a wide variety of useful applications especially for robot vision systems. By utilizing the capacity of the SOS, we are developing an intelligent electric wheelchair.

Figure 8.12 gives an outline of the system. Omnidirectional color images and range data acquired by the SOS are transmitted to a remote support person and/or a control PC (installed in the electric wheelchair) equipped with an automatic danger avoidance function.

In addition to being able to monitor the scene via a mobile phone or other means, the conveyed omni-directional images can be observed in an immersive display system to achieve a "telepresence" effect [10], which makes the remote support person feel present at the scene.

The automatic danger avoidance system uses omni-directional range information to detect obstacles and ground surface irregularities in real time, and then effects control to reduce speed or stop the wheelchair, as necessary. This kind of automatic danger avoidance function can serve to reduce the burden on the remote support person. For example, it makes conceivable a model in which one remote support person is able to support multiple electric wheelchair users.

8.5.2 System Configuration

Figure 8.13 shows the external appearance of the prototype intelligent electric wheelchair. Figure 8.14 shows omni-directional images rendered on a spherical

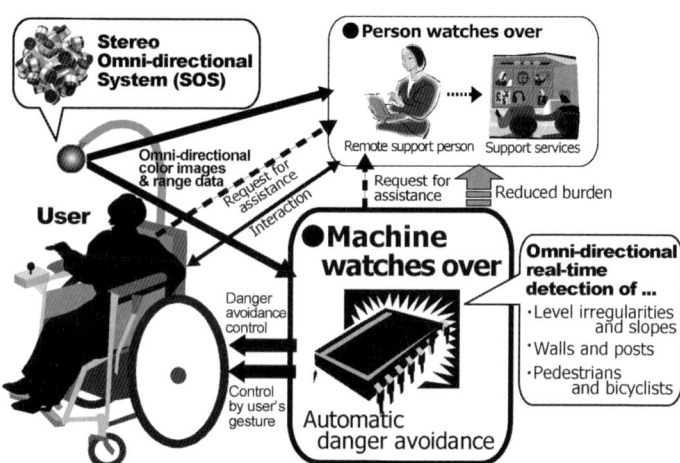

Fig. 8.12 The SOS supports safety of the electric wheelchair.

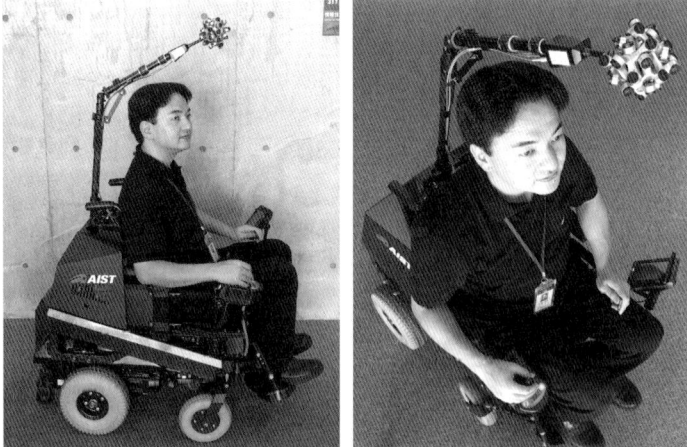

Fig. 8.13 External view of the prototype intelligent electric wheelchair.

Fig. 8.14 Spherical view of omni-directional images provided by the electric wheelchair.

surface using OpenGL. Figure 8.15 presents a block diagram explaining the system. The positioning of the SOS above and forward of the user's head offers the following advantages: (1) a wide observation range can be provided, without blind spots in the electric wheelchair's surroundings; (2) there is no hindrance to getting in and out of the wheelchair; and (3) the SOS is positioned at approximately body height (approx. 150 cm), thereby providing clearance in normal living spaces.

Integrated control of the SOS and electric wheelchair is handled by an on-board PC. The SOS and PC are connected by two 1.2 Gbps fiber optic lines, while control connection between PC and electric wheelchair is via RS232C. In addition to being used for obstacle avoidance and other automatic electric wheelchair control functions, the omni-directional data from the SOS can also be transmitted to remote locations over a wireless network.

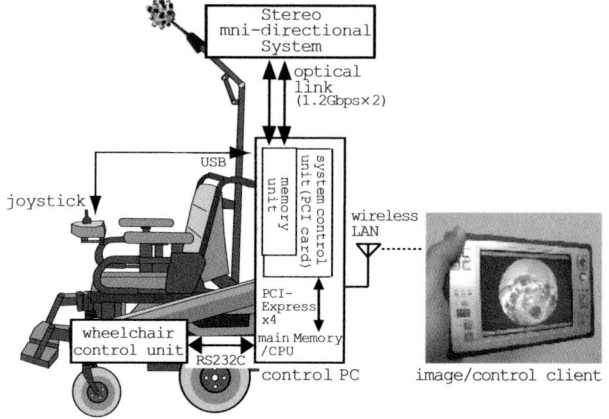

Fig. 8.15 Block diagram of the system.

Fig. 8.16 Omnidirectional depth image (bottom).

8.5.3 Obstacle Detection

Here we describe the function for automatic detection and avoidance of dangers that tend to occur commonly and frequently, such as impacts with obstacles and tipping over due to level irregularities. In this system, danger factors in the traveling environment are detected from the omni-directional range data supplied from the SOS. The detection objects considered in this study are as follows: (1) travel obstacles: obstacles on the ground surface, level irregularities, etc; (2) collision objects: pedestrians, walls, desks, etc; (3) other: ropes or beams suspended in the air, etc.

Figure 8.16 shows an omni-directional depth image actually captured by the SOS. The higher the brightness of the pixels, the shorter the range is. Figure 8.17 is

Fig. 8.17 Histogram plotting vertical frequency of the range data.

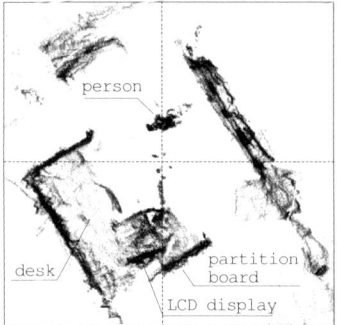

Fig. 8.18 Histogram representing an orthogonal projection of the range data onto the floor surface.

a histogram plotting vertical frequency of the range data. The vertical axis maps the vertical height, relative to the camera center (=0), and the horizontal axis maps the frequency. Since the SOS always measures uniformly in all directions, the range histogram accurately expresses the structural features of a room; we can see that large peaks are obtained at the positions corresponding to the ceiling and floor. (Since objects (e.g., desks) are present on the floor, the size of the peak for the floor is smaller than that for the ceiling.) Using this peak, the system automatically obtains the floor surface and ceiling reference position (only the floor surface in the case of outdoors). Figure 8.18 is a histogram representing an orthogonal projection of the range data onto the floor surface. This shows how objects that can potentially obstruct the movement of the electric wheelchair (center) can be detected. As the wheelchair is in actual movement, the histogram is used to narrow down the possible region of travel. Subsequently, the system uses the previously obtained information on the range to the ground to carefully examine the obstacles and level irregularities on the ground surface. Furthermore, in order to detect objects that are difficult to express in a histogram, such as a rope or beam suspended in the air, the system considers the 3-D position of edges that are near the wheelchair in the direction of travel. Note also that, in consideration of the user's intentions, the system will, in principle, effect automatic avoidance only by reducing speed or stopping the wheelchair.

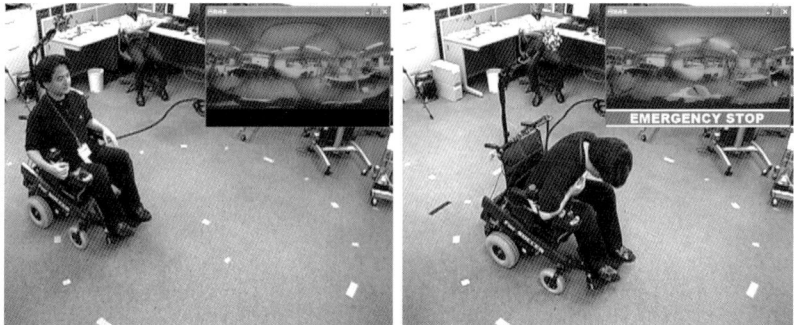

Fig. 8.19 Detection of user's behavior.

8.5.4 Gesture / Posture Detection

In this system, the SOS is positioned above and in front of the head of the user. Thus, while the environment surrounding the electric wheelchair is sensed, as described above, the user's riding posture and gestures can also be detected. Figure 8.19 shows an example of the system detecting an abnormality in riding posture (the user is slumped), and effecting an emergency stop of the wheelchair. Since during ordinary travel the upper body of the user remains essentially static, when the system detects any large change in posture it judges that an abnormality has arisen in the condition of the user, and thus activates an emergency stop. To detect the posture of the user, detailed position and shape information can be obtained, using the range information obtainable from the SOS. From this, the concept can be expanded to the detection of more intentional gestures such as finger-pointing gestures.

8.6 Conclusions

We have introduced a novel camera system capable of capturing omni-directional color images and range data simultaneously and in real time with a complete spherical field of view. Since the capability of the SOS is extremely powerful, it is expected to have a wide variety of useful applications. In this paper, we introduced an application to an intelligent electric wheelchair as the example. We also plan to consider another application such as for video surveillance systems.

References

1. Y. Yagi et al.: Real-time Omnidirectional Image Sensor (Copis) for Vision-Guided Navigation. IEEE Trans. on Robotics and Automation, **10**, pp. 11–22 (1994)
2. N. Winters et al.: Omnidirectional Vision for Robot Navigation. Proc. of IEEE Workshop on Omnidirectional Vision, pp. 21–28 (2000)

3. A.A. Argyros et al.: Robot Homing by Exploiting Panoramic Vision. Autonomous Robots 19, Springer Science + Business Media, pp. 7–25 (2005)

4. H. Ishiguro et al.: Omnidirectional stereo. IEEE Trans. on Pattern Analysis and Machine Intelligence. **14(2)**, pp. 257–262 (1992)

5. J. Gluckman and S.K. Nayar: Ego-Motion and Omnidirectional Cameras. Proc. of International Conference of Computer Vision, pp. 999–1005 (1998)

6. S. Nayar: Catadioptric omnidirectional camera. Proc. of Computer Vision and Pattern Recognition, pp. 482–488 (1997)

7. H. Tanahashi et al.: Acquisition of Three-Dimensional Information in Real Environment By Using Stereo Omnidirectional System(SOS). Proc. IEEE International Conference on 3D Digital Imaging and Modeling (3DIM2001), pp. 365–371 (2001)

8. S. Shimizu et al.: Moving object detection by mobile Stereo Omnidirectional System (SOS) using spherical depth image. Pattern Analysis & Applications, **9(2)**, pp. 113–126 (2005)

9. C. Wang et al.: Generation of Rotation Invariant Image Using Stereo Omnidirectional System (SOS). Proc. of the 10th Int. Conference on Virtual Systems and Multimedia (VSMM2004), pp. 269–272 (2004)

10. S. Moezzi (ed.): Special issue on immersive telepresence. IEEE MultiMedia **4(1)**, pp. 17–56. (1997)

Chapter 9
Video Analysis for Ambient Intelligence in Urban Environments

Andrea Prati and Rita Cucchiara

Abstract Ambient Intelligence (AmI) is an emerging field of research that comprises new paradigms, techniques and systems for intelligent processing of distributed sensing. A challenging arena for AmI framework is represented by urban environments that are characterized by high complexity, numerous sources of data, and spreading of interesting and non-trivial applications. In this context, the project LAICA (Laboratory of Ambient Intelligence for a Friendly City) represents a real experiment on the usefulness of AmI for advanced services for citizens. This chapter will address solutions of video analysis that can be directly applied in urban AmI. It describes in detail the uniqueness of LAICA approach, focusing in particular on the use of computer vision techniques for monitoring public parks. People surveillance and Web-based video broadcasting will be taken into account.

9.1 Introduction

With the term "Ambient Intelligence" (or AmI) we typically refer to new paradigms, techniques and systems for acquiring data, processing information, and creating and spreading knowledge in distributed environments. This new multi-disciplinary field of research has spread in the scientific community in recent decades, also thanks to the diffusion of sensors and the increase of processing power.

The typical contexts are areas where more heterogeneous sources of data coexist and can share raw and processed data. This sharing/cooperation among sensors

Andrea Prati

Dipartimento di Scienze e Metodi dell'Ingegneria, University of Modena and Reggio Emilia, Reggio Emilia, Italy, e-mail: andrea.prati@unimore.it

Rita Cucchiara

Dipartimento di Ingegneria dell'Informazione, University of Modena and Reggio Emilia, Modena, Italy, e-mail: rita.cucchiara@unimore.it

D. Monekosso et al. (eds.), *Intelligent Environments*, Advanced Information and Knowledge Processing, DOI: 10.1007/978-1-84800-346-0_9,

contributes to the common scope of improving the *"intelligence"*, with the Latin meaning of the act of *"intelligere"*, i.e., to *comprehend the world*.

Ambient intelligence research can be applied in the house, to improve the processing capabilities of past generations of home automation systems [15, 16], in distributed virtual communities for data sharing with autonomous mobile agents [33, 34, 35], in complex systems, such as the interaction between remote patients and health care systems [32]. Among the possible applications, one of the most challenging is the urban context, i.e., the city as a complex entity with people and numerous sources of data. The regional project LAICA (acronym for Laboratorio di Ambient Intelligence per una Città Amica - in English: Laboratory of Ambient Intelligence for a Friendly City) has been conceived in this framework and will be detailed in Section 9.2.1.

Among the many sources of information, videos assume a central role for many reasons. First, visual data can now be acquired at reasonable costs by using cheap cameras already installed in many public places (train stations, intersections, public parks, airports, etc.). Second, visual data are now much easier to transfer due to distributed wired and wireless networks available in most of the cities. Last but not the less important, visual data contain the highest amount of information about the environment and people who live in it.

For all these reasons, within LAICA we investigated the exploitation of visual data to extract information on status, behavior, and interaction of people and vehicles in urban contexts. Moreover, privacy and ethical issues will be taken into account and examples of applications in the LAICA projects will be described.

9.2 Visual Data for Urban AmI

In the past decades, visual documents have been the principal media of communication for tourist (virtual guides) and planning (remote sensing, SAR images) purposes for our cities. Nowadays, instead, the principal sources of visual information are live visual data acquired in real time from the hundreds of webcams and often cameras installed everywhere. Most of these cameras were installed only for tourist purposes (such as in the case of webpages of Times Square in New York City[1] - (USA) - or in Graz[2] - (Austria) - in which also PTZ control is made available to the user). However, also the thousands of cameras mounted as part of video surveillance systems can be potentially used.

The expression *video surveillance* is now synonymous of whichever system that uses cameras, acquires videos, possibly - but not necessarily - processes them, transfers them to remote displays in control centers and stores the data for posterity logging. However, existing video surveillance systems are constrained by strict privacy laws.

---

[1] http://www.earthcam.com/usa/newyork/timessquare/
[2] http://www.graz.at/cms/ziel/1097909/DE/

9.2.1 Video Surveillance in Urban Environment

Video surveillance is motivated by three main purposes, also known as the S^3 *motivations*: security, safety, statistics.

After the terrorist acts of September 11, 2001, every city in the world became insecure and the *possibility* to add "electronic eyes" to control the urban environment became an unavoidable *requirement*. This requirement reflected on the spreading of video surveillance systems in public places (especially metro and train stations, and airports) in order to prevent crimes, vandalism, or even terrorist attacks.

In Modena (Italy) the project "Modena Secure City" consisted of installing 42 cameras in critical locations, connected to the police control center and equipped with PTZ (Pan Tilt Zoom) capabilities to allow active control and zooming. Stored videos are available for forensic analysis. In the city of Reggio Emilia (Italy), before the start of the LAICA project, more than 100 cameras had been installed near the railway station and in several public parks. Other examples can be found all over the world, like New York City which has about 5000 cameras in Manhattan, or London which is the city with the highest number of cameras (approximately 150,000 in 2004) in the world. In total, Italy counted about 2 million cameras in 2004, while the UK reached, in the same year, the considerable count of 4 million with a citizen's picture taken on average 300 times per day.

The European commission has also expressed much interest in video surveillance in urban environments, since the Fifth Framework Programme (e.g., the Urbaneye Project[3], or CAVIAR project[4] in Sixth FP), and still much interest will be devoted to this research in the task of security in the Seventh Framework Programme, where a specific strategic objective (among others) called "Intelligent urban environment observation system" is included.

Finally, many commercial systems have been developed in recent years, some of them rather sophisticated, for instance the system developed by Bosch Security Systems[5], or the iOmniscient[6] (Australia) company that claims to have the most intelligent video surveillance system.

Video surveillance is becoming more and more popular for private use, in houses, offices, banks, to guarantee the safety of citizens and workers. New generations of video surveillance systems have also been mounted on mobile platforms. An example of this is the system developed by ELSAG (Italy) to automatically read vehicle license plates of stolen cars while a police car is moving. As an example of safety application, a system for smart deployment of airbags in the car has been developed at UCSD (USA) by the group of Mohan Trivedi [8]: here, multiple cameras (both standard and omnidirectional) are used to detect the driver's posture in real time in order to decide whether to deploy the airbag or not.

---

[3] http://www.urbaneye.net/index.html

[4] http://homepages.inf.ed.ac.uk/rbf/CAVIAR/caviar.htm

[5] http://www.boschsecurity.com

[6] http://iomniscient.com/

Finally, video surveillance can also be used for collecting statistics on people, behaviors, vehicles, etc.. These statistics can be used for dissemination to citizens or public officers, or for planning purposes. This is often the purpose of vision-based traffic monitoring systems, for both urban roads and highways, employed to measure queues, quantify lane occupancy and turning rates, detect incidents, measure speed, for access control to restricted areas, and so on. An interesting application related to traffic control and part of the LAICA project is that of monitoring roundabouts for occupancy analysis and license plate recognition.

For instance, the Belgian company Traficon N.V. is one of the world's leader in vision-based traffic control systems and has collaborated with us for more than two years on the development of a board-based system for safety in highway tunnels (called VIP-T) capable of detecting automatically incidents, monitoring vehicle flows, and collecting statistical data. Computer vision algorithms have been developed for vehicle detection and tracking with the purpose to measure speeds, classify vehicles, and detect stopped vehicles inside tunnels [11].

Statistical analysis is going to be of interest also when related to people, for example to analyze crowds of people in public places. Examples of applications are the monitoring of bus stops to plan frequencies of bus runs, crowd dynamics in department stores, etc. In all these cases privacy is a very relevant issue, because people are monitored without their explicit consent. This issue will be discussed in depth in Section 9.4.

As a final, emerging application, it is worth mentioning the use of video analysis for posterity logging to postprocess huge amounts of data for supporting forensic investigation in cases of crimes, vandalism, or terrorist attacks. For example, following the assassination of Professor Marco Biagi in Italy in 2002 more than 50,000 hours of videotapes have been watched and manually annotated by police officers. Having a (semi)automatic process for preprocessing these data would definitely help in such situations.

Summarizing, possible applications of video analysis in urban environments are reported in Table 9.1 and examples shown in Figure 9.1.

Table 9.1 Examples of applications of video analysis in urban environments.

Application	Example of system features
Traffic control	Queue analysis, incident detection, traffic light control
Monitoring and diffusing general information	Webcam broadcasting
People detection for safety purposes	Secure road crossings, metro/bus stop control
People tracking for security and surveillance	Surveillance of public areas (stadium, museum, etc.) or, in general, crowd areas
Environmental condition analysis	Fire and smoke detection, flood control
Citizen-to-computer interactions	Video interaction and communication systems (e.g., Infopoints)
Support for investigation	Posterity analysis for forensic purposes
Security for children	In the surroundings of schools or public parks, also in connection with the soliciting of minors
Control systems for cultural heritage	Monitoring of natural and historical parks, ...
Safety for elderly and children	Remote assistance for monitoring patients in intensive care units or quarantined patients

(a) Traffic control   (b) People tracking

(c) Web diffusion

Fig. 9.1 Some snapshots of possible applications of video surveillance in urban environments.

Video surveillance systems have greatly improved in recent decades. As reported in [25], broadly there are three generations of video surveillance systems. The first generation (up to 1980) was based on analog signal; videos were viewed remotely by human operators by means of a large set of monitors. These systems have the huge limitation of requiring the operators' attention, resulting in a high miss rate of the events of interest. Moreover, analog signal is very noisy and requires much bandwidth to be transmitted and much space to be stored. Thanks to the rapid improvements in camera resolution, the availability of low-cost computers, and communications improvements, in late 1980s, the second-generation systems began to emerge. These systems benefited from early advances in digital video communications (e.g., digital compression, bandwidth reduction, and robust transmission) and in computer vision algorithms, and were mainly used to show the feasibility of digital, intelligent attention focusing systems on video from limited sets of cameras and for real-time analysis and segmentation of image sequences, identification and tracking of multiple objects in complex scenes, human behavior understanding, etc. At the beginning of the 21st century, the third generation of video surveillance systems came about, providing a "full digital" solution to the design of surveillance systems: sensor and local processing layers can be physically organized together in a so-called intelligent camera; at the operator layer, an active interface is presented to the operator, assisting the operator by focusing his/her attention on a subset of interesting events.

Despite this classification, most of the existing video surveillance systems provide limited automated processing capability, by incorporating motion detectors for automatic storage of videos or a few more features. The goal of automated video surveillance is to extract meaningful objects from the observed scene, recognize them and their behavior, understand the scene by reasoning about objects and background, and infer specific conditions, alarms or interesting events.

A very promising advance to be included in the next (commercial) systems can be provided by including computer vision capabilities to the system. New types of alarms (not automatically provided in the majority of the current systems) could be:

- Low-level alarms: motion detectors, long-term change detectors, ...;
- Feature-based spatial alarms: specific-object detection in monitored areas (e.g., unattended bags in airports);
- Behavior-related alarms: anomalous trajectories, suspicious behaviors, ...;
- Complex event alarms: detection of complex scenarios related to multiple events.

9.2.2 The LAICA Project

The project LAICA is an example where many of the above-mentioned advanced capabilities have been tested in a distributed environment.

LAICA is a two-year (2005–2006) project funded by Regione Emilia-Romagna for a total budget of over 2 million euros and that involves universities, industries and public administrations for a total of about 320 man-months. The main objective of the LAICA project is to explore the AmI capabilities in a medium-size Italian city such as Reggio Emilia. LAICA partners aim at defining innovative models and technologies for AmI in urban environments, and at studying and developing advanced services for the citizens and the public officers in order to improve personal safety and prevent crimes. The project brings together the academic expertise and the industrial knowledge from several fields, ranging from low-power sensor networks, to computer vision, to middleware and mobile agents, and communication. Multimedia and multimodal data have been collected from different sources, such as cameras, microphones, and textual data about the traffic, security and the general situation of the city. As shown in Figure 9.2, the LAICA project has a three-layer architecture, corresponding to three different levels of granularity of the knowledge provided by sensors (punctual, local, and global). The processed information has been made available to both police control centers and citizens by means of a dedicated webpage.

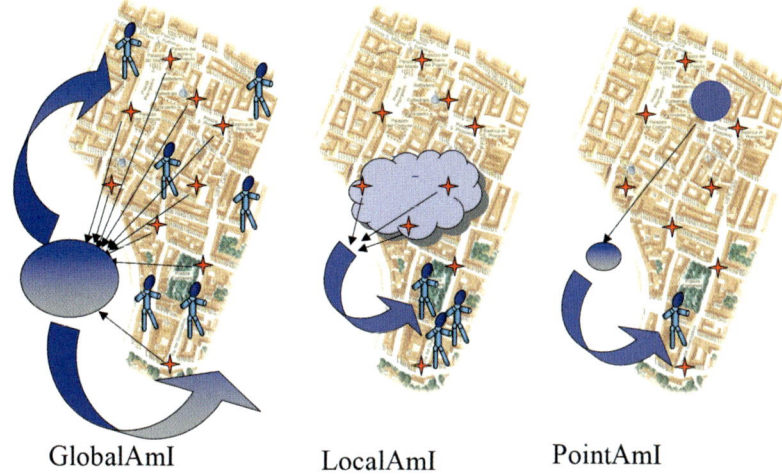

GlobalAmI   LocalAmI   PointAmI

Fig. 9.2 The three layers of the LAICA AmI architecture.

The foreseen services should be provided by a set of prototypical systems, as for instance:

- a system for the automatic monitoring of pedestrian subways by means of mobile and low-power audio and proximity sensors [45];
- a system for the automatic monitoring of traffic scenes by cameras for data collection and Web-based delivery of traffic news to citizens and police officers;
- a system that generates a feedback in pedestrian crossing systems to select the best duration of the green signal for the crossing [4];
- a platform of Urban TV to broadcast in interactive ways the data to the citizens;
- a system for the automatic monitoring of public parks with a plethora of cameras (both fixed and PTZ) [5, 41], also accounting for privacy issue [12].

In the next Section the last prototype will be discussed in detail to show how people can be detected and tracked in urban environments.

9.3 Automatic Video Processing for People Tracking

The motivations for tracking people in surveillance applications are numerous. The following is a list of the most important:

- Recognition of Human Motion, e.g.:

  - tracking people for statistical and security reasons, detecting moving people in dangerous zones;
  - walking, gait recognition;

- counting, locating pedestrians;
- abandoning an object;

• Gesture Recognition, e.g.:

- hand, arm tracking for gestures;
- sign language recognition;

• Tracking Faces in Video, e.g.:

- face detection;
- eye tracking and gaze tracking;
- lip reading, lip tracking;
- face recognition.

Most of these applications are particularly relevant in urban environments. Moreover, the city areas under surveillance are typically large, requiring multiple cameras to cover them. Finally, PTZ cameras are often employed to either "patrol" a large scene or zoom onto a specific zone/target. With these premises, the following subsections will briefly report on the research activity in the field of people detection and tracking by means of computer vision, starting from a single static camera, to multiple (static) cameras, to the use of PTZ cameras.

9.3.1 People Detection and Tracking from Single Static Camera

Detection of "moving objects" in video scenes is the basic step of major applications such as tracking and visual surveillance. This problem has been tackled for many years in both the scientific literature and the R&D for commercial systems and good solutions have already been proposed for static cameras. Among the many different approaches proposed, *background suppression* is the most used for its generality and reliability. The aim is to separate the foreground (moving visual objects, or MVOs) from the background model, i.e., a (probabilistic) model of the background as it changes in time. Thus, it is required to build and keep updated the background model, to *adapt* it to short- and long-term changes in illumination, to detect MVOs in the current frame (i.e., to suppress the background from it), and to handle difficult situations, such as shadows and the so-called "ghosts" (i.e., the false objects generated by a real still object that starts to move).

According to this, Table 9.2 shows a sketchy summary of these features (namely, background model construction, adaptive updating, suppression, and detection of other types of objects) and the most relevant approaches for background suppression used in the literature.

We also proposed an approach for background suppression from a single static camera. The approach is part of the SAKBOT (Statistical And Knowledge-Based Object Tracker) described with full details in [9]. The background pixels are defined by two models: the first statistical model updates the pixels at each frame using the

Table 9.2 Summary of seminal approaches to background suppression

Feature	Approach used
Background model construction	Median [19, 9]
	Single Gaussian [43, 26]
	Mixture of Gaussians [38]
	Eigenbackground [29]
Adaptive updating	Kalman-based [23]
	Previous backgrounds [43, 19]
Background suppression	Intensity [23]
	Color [19, 26, 9]
	Malahanobis distance [43]
	Multi-valued distance [38]
	Eigenbackground distance [29]
Detection of other objects	Shadows [26, 9], Ghosts [38, 9]

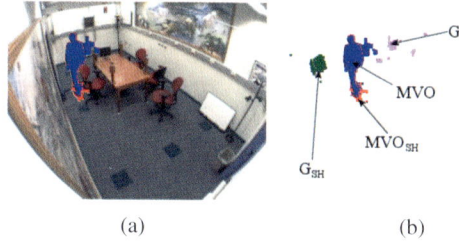
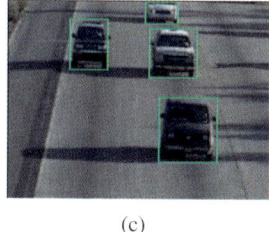

(a)   (b)   (c)

Fig. 9.3 Examples of the SAKBOT output: (a) segmentation result with blue pixels indicating the MVO, and red ones indicating the shadow points; (b) complete classification of pixels for frame (a) with MVO, MVO shadow, Ghost, and Ghost shadow; (c) another example of SAKBOT's segmentation (with corresponding bounding boxes).

temporal median function over the previous n sampled pixels; the second model exploits the knowledge of previous background and of the corresponding moving objects. Specifically, the pixels belonging to the current moving objects are not used for updating the model in order to prevent the gradual inclusion of slowly moving objects into the background. Instead, pixels detected as foreground at previous steps but classified as noise or shadows are included in the statistical model. This approach is critical when a stopped object starts to move, generating two "foreground" objects, one real and one apparent (the "ghost"). To avoid deadlock situations for the ghosts, a specific ghost suppression algorithm has been conceived. Moving shadows are classified using their appearance and assuming that shadows lower the brightness of the background underlying them, leaving the color components almost unchanged. Further details can be found in [31]. Figure 9.3 shows some examples of the output of the SAKBOT system.

Tracking of MVOs is crucial for most of the applications reported above. "Tracking" means that the same label/identity is kept constant over time for the same object, allowing computation of trajectory, speed, and behavior analysis. People tracking is one of the most widely explored topics in computer vision. There are many surveys in the field: the works of Cedras and Shah [7], of Gavrila [17], of Aggarwal and Cai [1] and Moeslund and Granum [28], or more recently, the work by Hu et al. in video surveillance [20] and the work by Wang et al. [42]. In people tracking, in order to cope with non-rigid body motion, frequent shape changes and self-occlusions, probabilistic and appearance-based tracking techniques are commonly proposed [26, 37]. In non-trivial situations, when several people interact overlapping each other, most of the basic techniques tend to lose the previously computed tracks, detecting instead the presence of a group of people, and possibly restoring the situation after the group has split up [26]. These methods aim at keeping the track history consistent before and after the occlusion only. Consequently, during an occlusion, no information about the appearance of the single person is available, limiting the efficacy of this solution in many cases. Conversely, a more challenging solution is to try to separate the group of people into individuals also during the occlusion.

Our approach to moving object tracking is based on appearance [40]. This algorithm uses a classical predict-update approach. It takes into account not only the status vector containing position and speed, but also the memory appearance model and the probabilistic mask of the shape [10]. The former is the adaptive update of each pixel in the color space. The latter is a mask whose values, ranging between 0 and 1, can be viewed as the probability for that pixel to belong to that object. These models are used to define a MAP (Maximum A Posteriori) classifier that searches the most probable position of each person in the scene. The tracking algorithm is a suitable modification of a work, previously proposed by Senior [37], that includes a specific module for coping with large and long lasting occlusions. Occlusions are classified into three categories: self-occlusions (or apparent occlusions), object occlusions, and people occlusions. Occlusion handling is very robust and has been tested in many applications. It can keep the shape of the tracked objects very precisely.

9.3.2 People Detection and Tracking from Distributed Cameras

The previous Section briefly described the relevant issues for detecting and tracking people from a single static camera. However, as reported above, in most of the urban scenarios a single camera does not suffice to handle large areas and complex/cluttered scenes. For this reason, multiple cameras are used to both provide multiple viewpoints (useful for disambiguating difficult situations by using redundant data and for handling occlusions) and obtain the coverage of a wider area. Unfortunately, in automatic video surveillance multiple cameras are useless if uncorrelated. The exploitation of the multiple viewpoints to correlate data from multiple cameras is often called *consistent labeling*, referring to the fact that the label/identity of moving objects is made consistent not only *over time* (as in the case of tracking from a single camera) but also *over space* (in the sense of different cameras). Consistent labeling permits tracking people in wide areas, increasing the potentiality of video-surveillance applications in urban scenarios.

Often cameras' fields of view are disjoint, due to installation and cost constraints. In this case, consistent labeling should be based on appearance only, basing the matching essentially on the color of the objects (such as color histogram matching [30]).

If the fields of view are overlapped, consistent labeling can exploit geometry-based computer vision. These approaches exploit geometrical relations and constraints between different views.

This could be done with a precise system calibration and 3D reconstruction could be used to solve any ambiguity [44]. However, this is not often feasible, in particular if the cameras are pre-installed and intrinsic and extrinsic parameters are not available. Thus, partial calibration or self-calibration methods can be adopted to extract only some of the geometrical constraints, e.g., to compute the ground-plane homography. An approach, based on the image projections of overlapped cameras' field of view lines, has been initially proposed by Khan and Shah in [22]: the lines delimiting the overlapping zones in the fields of view of the cameras are computed in a training phase with a single person moving in the scene. At run time, when one or more people have a camera hand-off, the distances from the lines are used to disambiguate objects, assuring label consistency.

Another class of approaches presented in the literature deals with multiview geometry to analyze and impose continuity in the objects' trajectory across camera streams (e.g., [2, 39]).

In [6], we have proposed a novel method, called *HECOL* (Homography and Epipolar-based COnsistent Labeling), to provide consistent labeling of people segmented in large areas covered by multiple overlapped cameras. The method takes into account both geometrical and shape features in a probabilistic framework. Homography and epipolar lines are computed to create relationships between cameras. The multi-camera system is modeled as a *Camera Transition Graph* (CTG) that defines the possible overlap between cameras in a given setup. When a new object is detected, the exploration of the graph selects a subset of compatible labels which may be assigned to the object in order to limit the search space. An off-line training phase allows computation of the *Entry Edges of Field of View* that define the area of overlapped FoV between cameras and permits construction of the homography. The learning phase also allows computing the location of the epipoles of the overlapped cameras with a robust algorithm based on RANSAC optimization.

At run time, the system checks for inconsistency in label assignments among the modules of the overlapped cameras. The main novelty of the paper lies in the phase of consistent labeling that defines a probabilistic framework with forward and backward contributions: it checks the mutual correspondence of people using the axis of the objects precisely warped in the other FoV using epipolar lines. It accounts for the matching of the warped axis and the shapes of people. This makes the method particularly robust against segmentation errors and allows to disambiguate groups of people. Figure 9.4 shows some images on the HECOL system.

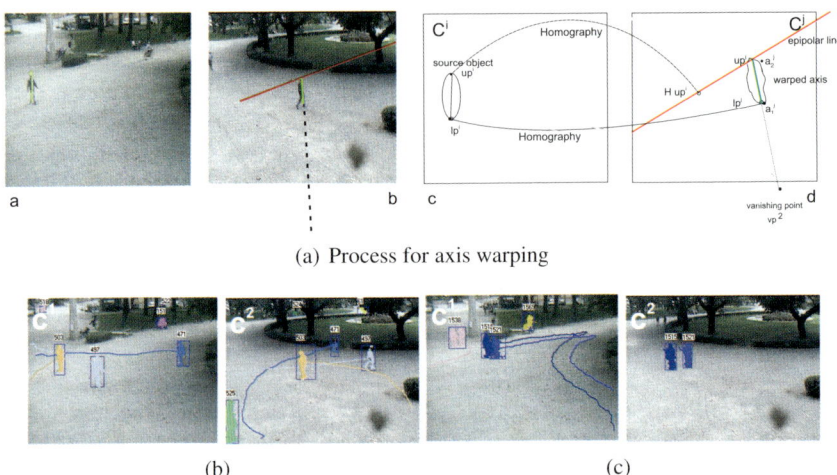

(a) Process for axis warping

(b)   (c)

Fig. 9.4 Examples of the HECOL system: (a) sketch of the process for axis warping on which the consistent labeling is based (see [6] for further details); (b) and (c) report examples from a real system working on a public park in Reggio Emilia within the LAICA project.

Multiple cameras can also be exploited directly to obtain 3D reconstruction of the moving object. In this case, the objective is not to have a consistent assignment of label among views but to correlate single points belonging to the object's shape, thus exploiting a sort of wide-baseline stereo system. An example of this use of multiple cameras can be found in [27]. The complete 3D reconstruction of the human shape has the obvious advantage of being crucial for applications such as human body modeling, gesture recognition, and similar.

9.3.3 People Detection and Tracking from Moving Cameras

In video-surveillance systems with multiple cameras it happens frequently that at least one of them is a PTZ camera. PTZ cameras have the main advantage of reducing the costs of covering large areas, allowing the use a single camera (even though more expensive than normal cameras), instead of a set of static cameras. PTZ camera can be programmed to patrol (automatically or manually by the operator) the scene. While patrolling, the camera should be able to extract and track moving people or even detect and track faces in order to zoom on one of them. Using a single PTZ camera solution has the advantages of being basically a cheaper solution, of not requiring synchronization/communication among cameras, and of requiring, in principle, lower computational load. It has, however, also some drawbacks, namely, the need for more complex computer vision techniques and the limitation of not allowing simultaneous coverage of a certain area.

Detecting and tracking people from a PTZ (i.e., moving) camera requires a rather different approach compared to what is reported in Section 9.3.1 for static cameras.

One possible approach is that of creating in real time a *mosaic image* of the whole scene (by registering overlapped images provided by successive frames of the active camera) then detecting and tracking moving people on the mosaic image.

The segmentation of moving objects becomes more critical when the video is acquired by a moving camera with an unconstrained and a priori unknown motion. Proposals from single camera can be grouped into three classes: based on *ego-motion computation*, on *motion segmentation*, and on *region merging with motion*. The approaches in the first class aim at estimating the camera motion (or ego-motion) through the evaluation of the dominant motion with different techniques and models in order to obtain compensated videos and to apply algorithms developed for fixed camera (frame differencing, as in [14], or background suppression, as in [36]). In [21] Kang et al. define an adaptive background model that takes into account the camera motion approximated with affine transformation. Tracking of moving object is achieved by means of a joint probability data association filter (JPDAF). In methods based on motion segmentation the objects are mainly segmented by using the motion vectors computed at pixel level [24]. The vectors are then clustered to segment objects with homogeneous motion. Finally, the approaches based on region merging with motion are hybrid approaches in which the objects are obtained with a segmentation based on visual features, and then merged on motion parameters computed at a region-level [18]. It is worth noting that most of the reported approaches are computationally very expensive and cannot meet real-time constraints (and those that meet them use either special-purpose devices or a set of limiting assumptions).

In [41] we proposed a new method for fast ego-motion computation based on the so-called *direction histograms*. The method works with an uncalibrated camera that moves with an unknown path and it is based on the compensation of the camera motion (i.e., the *ego-motion*) to create the mosaic image and on the frame differencing to extract moving objects. Successive steps eliminate the noise and extract the complete shape of the moving objects in order to exploit an appearance-based probabilistic tracking algorithm. Figure 9.5 shows an example of the segmentation of a moving person by means of a single PTZ camera and its exploitation for automatically following the person.

9.4 Privacy and Ethical Issues

All the considerations reported in previous Sections are related to the usefulness of video surveillance in urban environments for increasing (the sense of) security, safety or for the collection of statistical information. These are obvious and undoubted advantages of (automatic) video surveillance. The use of cameras in public places creates, however, serious problems regarding the citizen privacy. There is a worldwide controversy regarding the use of video surveillance in public places. This has to do with privacy violations. The dichotomy security vs. privacy was and is, for instance, very debated in United States after September 11. K.W. Bowyer wrote a

Fig. 9.5 Examples of the detection of a single moving person by using a single PTZ camera. Additionally, PTZ camera moves automatically to follow the moving person.

very interesting paper on pros and cons of surveillance and analyzed the risks of false claims in privacy violations [3].

In Europe, the privacy debate has been magnified following the terrorist incidents in Madrid in March 2004 and on the London underground in July 2005. In both cases, the recorded videos provided valuable help for the identification of the terrorists only after the crime.

In Europe visual data processing is within a more general Directive (95/46/EC) than in United States. This Directive covers specific features of the processing of personal information included in sound and image data and ensures "the protection of privacy and private life as well as the larger gamut of protection of personal data with regard to fundamental rights and freedoms of natural persons". A considerable portion of the information collected by means of video surveillance concerns identified and/or identifiable persons, who have been filmed as they moved in public and/or publicly accessible premises. As a final remark, the directive states that "*in public places no automatic visual surveillance should limit the freedom of people*".

Each European country has then its own specific law. For instance, in Denmark surveillance of public streets, roads, squares or any similar area used for common travel is forbidden to private entities. Also in Italy, there exists a specific set of laws for video surveillance. These laws propose a basic principle called "proportionality principle", for which acquired data must be adequate, relevant and not excessive. As

Fig. 9.6 Example of face obscuration taken from [12].

an example, acquiring and storing videos from a supermarket for statistical analysis is excessive, doing that for forensic analysis of crimes is not.

A good compromise between security and privacy comes from the use of computer vision. It allows the extraction of "biometric" information (such as faces) from the video, but still preserving semantic content to be freely distributed. This requires, as depicted in previous Sections, to detecting and track people from multiple cameras, detect their faces and automatically obscuring them to prevent "identification" of the person.

In the framework of the project LAICA we have studied and developed two different solutions: the first makes use of passive sensors to develop a video-surveillance system integrated with the cameras [13], the second automatically extracts and obscures people's faces from videos [12]. An example of face obscuration is given in Figure 9.6.

References

1. Aggarwal, J. K., Cai, Q.: Human Motion Analysis: A Review. Computer Vision and Image Understanding. **73(3)**, pp. 428–440 (1999)
2. Black, J., Ellis, T.: Multi camera image tracking. Image and Vision Computing. **24(11)**, pp. 1256–1267 (2006)
3. Bowyer, K.W.: Face recognition technology and the security versus privacy tradeoff. IEEE Technology and Society. **1**, pp. 9-20 (2004)
4. Broggi, A., Fedriga, R.I., Tagliati, A., Graf, T., Meinecke, M.: Pedestrian Detection on a Moving Vehicle: an Investigation about Near Infra-Red Images. In: Proceedings of IEEE Intelligent Vehicle Symposium (IV), pp. 431–436 (2006)
5. Calderara, S., Cucchiara, R., Prati, A.: Group Detection at Camera Handoff for Collecting People Appearance in Multi-camera Systems. In: Proceedings of Conference on Advanced Video and Signal-based Surveillance (IEEE AVSS 2006), pp. 36–41 (2006)
6. Calderara, S., Prati, A., Cucchiara, R.: HECOL: Homography and Epipolar-based Consistent Labeling for Outdoor Park Surveillance. Computer Vision and Image Understanding (2007)
7. Cedras, C., Shah, M.: Motion-Based Recognition: A Survey. Image and Vision Computing. **13(2)** (1995)
8. Cheng, S.Y., Trivedi, M.M.: Human posture estimation using voxel data for "smart" airbag systems: issues and framework. In: Proceedings of IEEE Intelligent Vehicles Symposium (IV), pp. 84–89 (2004)

9. Cucchiara, R., Grana, C., Piccardi, M., Prati, A.: Detecting Moving Objects, Ghosts and Shadows in Video Streams. IEEE Transactions on Pattern Analysis and Machine Intelligence. **25(10)**, pp. 1337–1342 (2003)

10. Cucchiara, R., Grana, C., Tardini, G., Vezzani, R.: Probabilistic People Tracking for Occlusion Handling. In: Proceedings of IAPR International Conference on Pattern Recognition (ICPR 2004), vol. 1, pp. 132–135 (2004)

11. Cucchiara, R., Melli, R., Prati, A., De Cock, L.: Predictive and Probabilistic Tracking to Detect Stopped Vehicles. In: Proceedings of Workshop on Applications of Computer Vision (WACV), pp. 388–393 (2005)

12. Cucchiara, R., Prati, A., Vezzani, R.: A System for Automatic Face Obscuration for Privacy Purposes. Pattern Recognition Letters. **27(15)**, 1809–1815 (2006)

13. Cucchiara, R., Prati, A., Vezzani, R., Benini, L., Farella, E., Zappi, P.: An Integrated Multi-Modal Sensor Network for Video Surveillance. Journal of Ubiquitous Computing and Intelligence (JUCI). **1**, pp. 1–11 (2007)

14. Cutler, R., Davis, L.S.: Robust real-time periodic motion detection. IEEE Transactions on Pattern Analysis and Machine Intelligence. **22(8)**, pp. 781–796 (2000)

15. Friedewald, M., Da Costa, O., Punie, Y., Alahuhta, P., Heinonen, S.: Perspectives of ambient intelligence in the home environment. Telematics and Informatics. **22(3)**, 221–238 (2005)

16. Garate, A., Lucas, I., Herrasti, N., Lopez, A.: Ambient intelligence as paradigm of a full automation process at home in a real application. In: Proceedings of IEEE International Symposium on Computational Intelligence in Robotics and Automation, CIRA, pp. 475–479 (2005)

17. Gavrila, D.M.: The Visual Analysis of Human Movement: A Survey. Computer Vision and Image Understanding. **73(1)**, pp. 82–98 (1999)

18. Gelgon, M., Bouthemy, P.: A region-level motion-based graph representation and labeling for tracking a spatial image partition. Pattern Recognition. textbf33, pp. 725–740 (2000)

19. Haritaoglu, I., Harwood, D., Davis, L.S.: W4: real-time surveillance of people and their activities. IEEE Transactions on Pattern Analysis and Machine Intelligence. **22(8)** (2000)

20. Hu, W., Tan, T., Wang, L., Maybank, S.: A survey on visual surveillance of object motion and behaviors. IEEE Transactions on Systems, Man, and Cybernetics - Part C. **34(3)**, pp. 334–352 (2004)

21. Kang, J., Cohen, I., Medioni, G.: Continuous tracking within and across camera streams. In: Proceedings of IEEE-CS Int'l Conf. on Computer Vision and Pattern Recognition (CVPR), vol. 1, pp. I-267 - I-272 (2003)

22. Khan, S., Shah, M.: Consistent labeling of tracked objects in multiple cameras with overlapping fields of view. IEEE Transactions on Pattern Analysis and Machine Intelligence. **25(10)**, pp. 1355–1360 (2003)

23. Koller, D., Weber, J., Huang, T., Malik, J., Ogasawara, G., Rao, B., Russel, S.: Towards Robust Automatic Traffic Scene Analysis in Real-Time. In: Proceedings of International Conference on Pattern Recognition (1994)

24. Lee, K.W., Ryu, S.W., Lee, S.J., Park, K.T.: Motion based object tracking with mobile camera. Electronic Letters. **34(3)**, pp. 256–258 (1998)

25. Marcenaro, L., Oberti, F., Foresti, G.L., Regazzoni, C.S.: Distributed architectures and logical-task decomposition in multimedia surveillance systems. Proceedings of the IEEE. **89(10)**, 1419–1440 (Oct. 2001)

26. McKenna, S.J., Jabri, S., Duric, Z., Rosenfeld, A., Wechsler, H.: Tracking groups of people. Computer Vision and Image Understanding. **80(1)** (2000)

27. Mikic, I., Trivedi, M.M., Hunter, E., Cosman, P.C.: Human Body Model Acquisition and Tracking Using Voxel Data. International Journal of Computer Vision. **53(3)** pp. 199–223 (2003)

28. Moeslund, T.B., Granum, E.: A Survey of Computer Vision-Based Human Motion Capture. Computer Vision and Image Understanding. **81**, pp. 231–268 (2001)

29. Oliver, N.M., Rosario, B., Pentland, A.P.: A Bayesian computer vision system for modeling human interactions. IEEE Transactions on Pattern Analysis and Machine Intelligence. **22(8)** (2000)

30. Orwell, J., Remagnino, P., Jones, G.A.: Multi-camera color tracking. In: Proceedings of Second IEEE Workshop on Visual Surveillance (VS'99), pp. 14–21 (1999)
31. Prati, A., Mikic, I., Trivedi, M.M., Cucchiara, R.: Detecting Moving Shadows: Algorithms and Evaluation. IEEE Transactions on Pattern Analysis and Machine Intelligence. **25(7)**, pp. 918–923 (2003)
32. Riva, G.: Ambient Intelligence in Health Care. Cyberpsychology and Behavior. **6(3)**, 295–300 (2003)
33. Riva, G., Davide, F., Ijsselsteijn, W.A.: Being There: Concepts, effects and measurements of user presence in synthetic environments. IOS Press (2003)
34. Satoh, I.: Software Agents for Ambient Intelligence. In: Proceedings of IEEE International Conference on Systems, Man and Cybernetics, pp.1147–1150 (2004)
35. Satoh, I.: Mobile Agents for Ambient Intelligence. In: Lecture Notes in Computer Science (LNCS), vol. 3446, Springer (2005)
36. Sawhney, H., Ayer, S.: Compact representations of videos through dominant and multiple motion estimation. IEEE Transactions on Pattern Analysis and Machine Intelligence. **18(8)**, pp. 814–830 (1996)
37. Senior, A.: Tracking people with probabilistic appearance models. In: Proceedings of Int'l Workshop on Performance Evaluation of Tracking and Surveillance (PETS) Systems, pp. 48–55 (2002)
38. Stauffer, C., Grimson, W.E.L.: Learning patterns of activity using real-time tracking. IEEE Transactions on Pattern Analysis and Machine Intelligence. **22(8)** (2000)
39. Tsutsui, H., Miura, J., Shirai, Y.: Optical Flow-based Person Tracking by Multiple Cameras. In: Proc. 2001 Int. Conf. on Multisensor Fusion and Integration in Intelligent Systems, pp. 91–96 (2001)
40. Vezzani, R.: Computer Vision for People Video Surveillance Ph.D. Thesis. (2006) Available via Internet. http://imagelab.ing.unimo.it/Pubblicazioni/publications_query.asp?lang=en\&autore=+55+\&categoria=0\&tipo=5.Cited12Aug2007
41. Vezzani, R., Prati, A., Cucchiara, R.: Advanced Video Surveillance with Pan Tilt Zoom Cameras. In: Proceedings of Workshop on Visual Surveillance (VS) (2006)
42. Wang, L., Hu, W., Tan, T.: Recent developments in human motion analysis. Pattern Recognition. **36(3)**, pp. 585–601 (2003)
43. Wren, C., Azarbayejani, A., Darrell, T., Pentland, A.P.: Pfinder: real-time tracking of the human body. IEEE Transactions on Pattern Analysis and Machine Intelligence. **19(7)** (1997)
44. Yue, Z., Zhou, S.K., Chellappa, R.: Robust two-camera tracking using homography. In: Proceedings of IEEE Intl Conf. on Acoustics, Speech, and Signal Processing, vol. 3, pp. 1–4 (2004)
45. Zappi, P., Farella, E., Benini, L.: A PIR based wireless sensor node prototype for surveillance applications. In: Proceedings of European Workshop on Wireless Sensor Networks (EWSN 06), pp. 26–27 (2006)

Chapter 10
From Monomodal to Multimodal: Affect Recognition Using Visual Modalities

Hatice Gunes and Massimo Piccardi

Abstract Affective computing has emerged with the aim to enable affective human-computer interaction by designing machines and interfaces that will sense, recognize, understand and interpret human emotional states via language, speech, facial and bodily gesture and respond accordingly. Although much progress has been achieved in the last decade, one major present limitation of affective computing has been that most of the research on emotion recognition has focused on one single sensorial source, or modality, at a time and especially the face display. While it is true that the face is the main display of a human's affective state, other sources can improve the recognition accuracy. As natural human-to-human interaction is multimodal, the single sensory observations are often ambiguous, uncertain, and incomplete. Despite this fact, the research community has only recently started proposing emotion recognition systems using affective multimodal data. This chapter introduces recent advances in multi-modal affect. It explicitly focuses on systems that include vision as one of the input modalities, and attempt to analyze affective face and body movement either as a pure monomodal system or as part of a bi-modal/multimodal affective framework introduced during the period 2002-2006.

10.1 Introduction

Research on automatic emotion recognition did not start until the 1990s. Although researchers like Ekman published studies on how people recognized emotions from face display in the 1960s [16], people would find it absurd that anyone would even propose giving machines such abilities when emotional mechanisms were not considered to have a significant role in various aspects of a human's life. However,

Faculty of Information Technology, University of Technology, Sydney (UTS)
P.O. Box 123, Broadway 2007, NSW, Australia
haticeg,massimo@it.uts.edu.au

D. Monekosso et al. (eds.), *Intelligent Environments*, Advanced Information
and Knowledge Processing, DOI: 10.1007/978-1-84800-346-0_10,
© Springer-Verlag London Limited 2009

scientists found out that even in the most rational of decisions, emotions persist: emotions always exist, we always feel something.

In the early 1990s, Mayer and Salovey published a series of papers on emotional intelligence [56]. They suggested that the capacity to perceive and understand emotions defines a new variable in personality. Goleman popularized his view of emotional intelligence or Emotional Quotient (EQ) in his 1995 best-selling book by discussing why EQ mattered more than Intelligence Quotient (IQ) [23]. Goleman drew together research in neurophysiology, psychology and cognitive science. Other scientists also provided evidence that emotions were tightly coupled with all functions that humans are engaged with: attention, perception, learning, reasoning, decision making, planning, action selection, memory storage and retrieval [31, 49].

This new scientific understanding of emotions provided inspiration to various researchers for building machines that will have abilities to recognize, express, model, communicate, and respond to emotions. The initial focus has been on the recognition of the prototypical emotions from posed visual input, namely, face expressions. All existing work in the early 1990s attempted to recognize prototypical emotions from two static face images: neutral and expressive. In the second half of the 1990s, automated face expression analysis started focusing on posed video sequences and exploiting temporal information in the displayed face expressions. In parallel to the automatic emotion recognition from visual input, works focusing on audio input emerged.

Rosalind Picard's award-winning book, *Affective Computing*, was published in 1997, laying the groundwork for giving machines the skills of emotional intelligence. The book triggered an explosion of interest in the emotional side of computers and their users and a new research area called *affective Computing* emerged. *Affective Computing* advocated the idea that it might not be essential for machines to possess all the emotional intelligence and skills humans do. However, for natural and effective human-computer interaction, computers still needed to look intelligent to some extent [50]. Experiments conducted by Reeves and Nass showed that for an intelligent interaction, the basic human-human issues should hold [54].

One major limitation of affective computing has been that most of the past research had focused on emotion recognition from one single sensorial source, or modality. However, as natural human-human interaction (HHI) is multimodal, the single sensory observations are often ambiguous, uncertain, and incomplete. It was not till 1998 that computer scientists attempted to use multiple modalities for recognition of emotions/affective states. The combined use of multiple modalities for sensing affective states in itself triggers another research area. What channels to use? And how to combine them? The initial interest was on fusing visual and audio data. The results were promising; using multiple modalities improved the overall recognition accuracy helping the systems function in a more efficient and reliable way. Starting from the work of Picard in 2001, interest in detecting emotions from physiological signals emerged. Moreover, researchers moved their focus from posed to spontaneous visual data [5]. Although a fundamental study by Ambady and Rosenthal suggested that the most significant channels for judging behavioral cues of humans appear to be the visual channels of face expressions and body gestures [2], the existing literature on automatic emotion recognition did not focus on the expressive information that body gestures carry till 2003 [30].

Although the most common approach has been that of combining face expressions with audio information, following the new findings in psychology, some researchers advocate that a reliable automatic affect recognition system should attempt to combine face expressions and body gestures. Accordingly, a limited number of approaches have been proposed for such sensorial sources [3], GUNES-Gun2006-1, GUNES-Kap2005, GUNES-Lis2002, GUNES-Mad2004. With all these new areas, a number of new challenges have arisen. The stage affective computing has reached today is combining multiple channels for affect recognition and moving from posed data toward spontaneous data. Achieving these aims is an open challenge. At this level, scientists expect emotion recognition to be solvable by machine in the near future, at least as well as people can label such patterns [49].

10.2 Organization of the Chapter

There is a vast body of literature on emotion recognition from individual modalities like face expression and audio signals. Here, instead of a comprehensive survey, we discuss work not included in previous surveys while highlighting the main research issues.

Currently, there are very few multimodal systems attempting to analyze combinations of communication means for human affective state analysis. There exist bimodal systems for affect recognition combining audio and video signals by processing face expression and vocal cues. Such systems have been reviewed elsewhere [47]. Examples of similar systems introduced from the year 2002 onwards are [8], [22] and [60].

In relation to what this chapter has set out to explore, we are interested in systems that attempt to analyze the nonverbal communication of emotions. We explicitly focus on systems that include vision as one of the input modalities, and attempt to analyze affective face and body movement either as a pure monomodal system or as part of a bimodal/multimodal affective framework introduced during the period 2002–2006. Facial expression recognition systems have been reviewed in [20] and [48]. Hence, we only briefly focus on the most recent research conducted in the area in the last five years. Automated systems analyzing and recognizing affective body movement first emerged in 2002. As these systems do not have a lengthy background similar to that in face expression analysis, the attempts are relatively few. The review of these systems naturally leads to exploration of the literature in the combination of modalities for emotion recognition. We thus describe the most notable multimodal systems combining face and body modalities for emotion recognition.

The rest of the chapter is organized as follows. Section 10.3 is concerned with the challenges faced when moving from monomodal affect recognition systems to multimodal ones and discusses the problem domain of multimodal affective computing.

Section 10.4 explores the systems recognizing affective face or body movement, either by focusing on the prototypical face expressions/face action units (AUs) or body expressions. Due to limitation in space, the chapter in general and this section in particular only briefly cover the representative systems, referring the reader to [24] for further details. Section 10.5 covers the systems recognizing affective bimodal/multimodal data from visual and/or haptic modalities. Representative systems are analyzed and compared. The last part of this chapter discusses the future of affective multimodal recognition systems, lists the limitations of the current systems and summarizes the features of an *ideal* multimodal affect analyzer.

10.3 From Monomodal to Multimodal: Changes and Challenges

This section is concerned with the challenges faced when moving from monomodal affect recognition systems to multimodal ones. We describe the new problems occurring and changes needed by this shift and propose some solutions. Functional blocks for an affect recognition system with a comparison of monomodal vs. multimodal are provided in Figure 10.1. By looking at the figure, one can note that some of the assumptions made when building monomodal affect recognizers still hold (e.g., affect data collection is still needed). However, specific problems exist for multimodal affect recognition (e.g., multiple sensors are now required). Therefore, some new assumptions need to be taken into consideration. We choose to focus specifically on the following challenges: background research, data collection, data annotation, synchrony between modalities, data integration/fusion, information complementarity/redundancy and information content of modalities. We discuss them in detail in the following subsections.

10.3.1 Background Research

As we already stated in the introduction of this section, development of affective multimodal systems depends significantly on the progress in emotion research. Such progress is likely to take place thanks to the interaction of researchers on human emotions and computer scientists.

People are challenged in daily life with the task of decoding and making sense of multiple simultaneously presented emotional signals. Emotional information is conveyed by various physical changes in the body: changes invisible to others (e.g., blood chemistry, brain activity, neurotransmitters) and/or physical changes that can be differentiated by humans (voice, tone, face, gesture) [38]. Hence, a broad range of modalities is available, including speech and language, gesture and head movement, body movement and posture, as well as face expression. Psychologists state that 93% of HHI and communication is nonverbal and humans display their emotions most expressively through face expressions and body gestures [41].

Computers can also measure affect that is clearly expressed to them. It is possible to measure face and body activities that might not be visible using electromyography

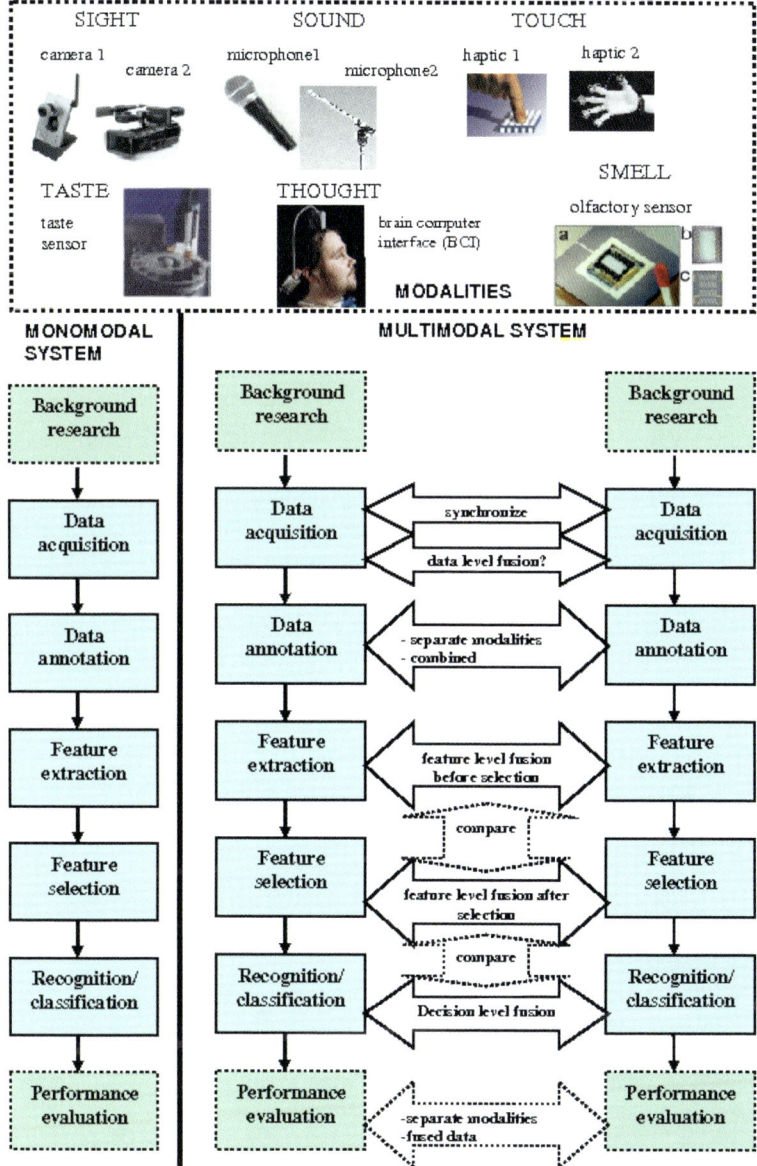

Fig. 10.1 Functional blocks for an affect recognition system: monomodal vs. multimodal.

(EMG) [38]. However, *visual communicative cues* are commonly accepted to be a social sign for their *signification of intent* as the way humans interact with each other. Moreover, within the *visual communicative cues* expressive face and body gesture are among the main nonverbal communication channels in HHI [2, 14, 41]. Hence, understanding human emotions through these nonverbal means is one of the necessary skills both for humans and computers to interact intelligently.

One limitation of prior work on human emotion perception is the focus on separate channels for expression of affect, without adequate consideration for the multimodal emotional signals that people encounter in their environment [59]. Most research on the development of emotion perception has focused on human recognition of face expressions, and thus we know little about the relative influence of other emotional expressions on human perceptual and attentional abilities. The investigation of various ways in which people learn to perceive and attend to emotions multimodally will likely provide a more complete picture of the complex HHI.

Herewith, we provide a summary of the findings from emotion research relevant to what this chapter has set out to explore: emotion communication from face and body display.

Face Expression

Ekman conducted various experiments on human judgment on still photographs of posed face behavior and concluded that seven basic emotions can be recognized universally: neutrality, happiness, sadness, surprise, fear, anger and disgust [18]. Several other emotions and many combinations of emotions have been studied but it remains unconfirmed whether they are universally distinguishable. Although prototypic expressions, like happiness, surprise and fear, are natural, they occur infrequently in daily life and provide an incomplete description of face expression. To capture the subtlety of human emotion and paralinguistic communication, automated recognition of fine-grained changes in face expression is needed. Ekman and Friesen developed their *Facial Action Coding System* (FACS) for describing face expressions by face action units (AUs) [17]. FACS is based on the enumeration of all "face action units" causing face movements (see Figure 10.2(a) for examples

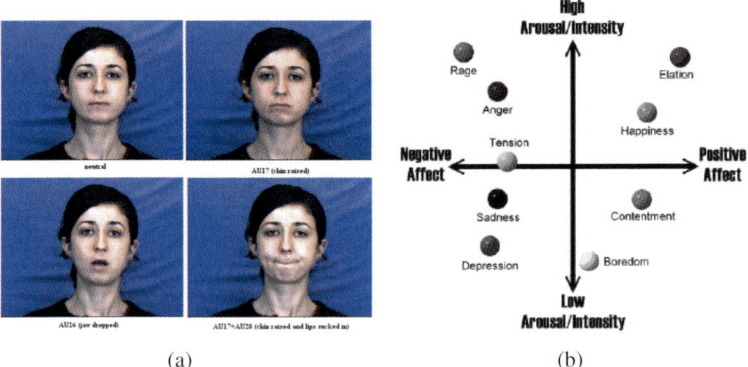

(a)   (b)

Fig. 10.2 (a) Examples of AU activations (from left to right: neutral; AU17: chin raised; AU26: jaw dropped; AU17+AU28: chin raised and lips sucked) and (b) representation of Russell's circumflex model (based on [55]).

of AUs). After an extensive analysis, Ekman and Friesen decided for a final number of 46 AUs which account for changes in face expression, 30 AUs anatomically related to the contractions of specific face muscles: 12 for the upper face and 18 for the lower face. A set of translation rules are used to link the AU coding into basic emotions. For instance, the presence of four AUs can be interpreted as the emotion "surprise" [17]: *AU1+ AU2+ AU5+AU26==Surprise* (AU1: Inner Brow Raised; AU2: outer brow raised; AU5: upper lid raised; AU26: jaw dropped). To date, Ekman's theory of emotion universality [18] and the Facial Action Coding System (FACS) [17] are the most commonly used schemes in vision-based systems attempting to recognize face expressions.

Body Expression

Compared to research in face expression, the expressive information body gestures carry has not been adequately exploited yet. However, the interest is growing. Neuroscientists conducted an experiment to determine the underlying neural mechanisms of perception of body expression of emotion [29]. Their findings suggest that the brain reacts as quickly and with the same pair of neural structures as it does in the case of face expressions, thus confirming the fact that body expression is an integral part in emotion communication [29]. Coulson presented experimental results on attribution of six emotions (anger, disgust, fear, happiness, sadness and surprise) to static body postures by using computer-generated figures [13]. He found out that in general, human recognition of emotion from posture is comparable to recognition from the voice, and some postures are recognized as effectively as face expressions. Moreover, a fundamental study by Ambady and Rosenthal suggested that the most significant channels for judging behavioral cues of humans appeared to be the visual channels of face expressions and body gestures [2].

In general, the body and hand gestures are much more varied than face gestures. There is an unlimited vocabulary of body postures and gestures with combinations of movements of various body parts. Despite the effort of Laban in analyzing and annotating body movement [36, 37] unlike the face action units, body action units that carry expressive information have not been defined with a Body Action Coding System (BACS). Communication of emotions by body gestures is still an unresolved area in psychology.

When recognizing and labeling affect data from body display, Russell's theory of arousal and valence [55] is commonly used. Russell viewed affective states not independent of one another; rather, related to one another in a systematic manner [55]. Russell argued that emotion is best characterized in terms of a small number of latent dimensions, rather than in a small number of discrete emotion categories. He proposed that each of the basic emotions is a bipolar entity as part of the same emotional continuum. The proposed polarities are arousal (relaxed vs. aroused) and valence (pleasant vs. unpleasant). The model is illustrated in Figure 10.2(b).

10.3.2 Data Collection

When shifting our focus from monomodal to multimodal affect recognition, databases containing representative samples of human *multimodal* expressive behaviors are needed for the development of such systems. Hence, the requirement now becomes that of databases containing data from *multiple* channels/sensors.

In order to describe the problem domain of multimodal affect data collection, we should first focus on the factors that affect the extent and the nature of this task. These factors were defined by Picard as follows [51]:

- *Posed/spontaneous:* Is the emotion elicited by the subject upon request or is there an actual reason or situation creating the affective activation?
- *Expression/emotion:* Is the actual target on expression (how people externalize) or emotion (what people feel internally)?
- *Laboratory setting/real life:* Is the recording obtained in a laboratory with controlled background/lights/noise or in real life with unconstrained conditions?
- *Open recording/hidden recording:* Does the subject know that s(he) is being recorded?
- *Emotion-purpose/other-purpose:* Does the subject know that s(he) is expected to create emotional response?

To foster development of natural human-computer interfaces, an ideal multimodal affect database should contain data obtained in a natural setup; in other words, data that are spontaneous and obtained in a real life situation with non-emotion purpose. Taking into account the aforementioned factors, an ideal multimodal affect database thus should have the following features:

- The subjects are present in their *natural environment* (e.g., office or house).
- The subjects are in a particular affective state due to some real-life event or trigger of events (e.g., stressed at work).
- The subjects *do not try to hide nor exaggerate* what they feel, on the contrary, display what they feel using multiple communicative channels (e.g., face expression, head movement, body gestures, voice, etc.).
- The subjects *are not aware of the recording*, hence will not restrain themselves unlike the case when they are part of an experiment; but know they can be seen (e.g., public/not alone).
- There are occurrences of *occlusions* (e.g., hands occluding each other or hand occluding the face) and noise (e.g., in audio recordings).
- There may be *multiple sensing devices* (e.g., multiple cameras, multiple microphones, haptic sensors, etc.).
- *Viewing and lighting conditions are realistic (e.g., with background noise) and not uniform.*
- The *sessions are long and varied*, expanding between one day and possibly a couple of weeks, capturing *all variations* of expressive1-expressive2-expressive3-neutral behavior in every possible order or combination.
- The subjects are of *diverse age, gender and ethnic background.*

At present publicly available databases exist mainly for single expressive modalities such as face expressions, static and dynamic hand postures, and dynamic hand gestures. Recently, there have been a limited number of attempts to create publicly available multimodal affect databases. These are: the SmartKom Corpora [1], FABO [25], the Database collected at the University of Amsterdam [58] and the Databases collected at the University of Texas [61]. These are reviewed in [26] in detail. To date, FABO is the first and only bimodal affect database consisting of expressive face and upper-body display [26].

In a multimodal affect database, when evaluating posed versus spontaneous data, artificial occurrences should be distinguished from natural occurrences [15]. As confirmed by many researchers in the field, directed affective face and body action tasks differ in appearance and timing from spontaneously occurring behavior [10]. Deliberate face and body behavior are mediated by separate motor pathways and differences between spontaneous and deliberate actions may be significant [32]. Very few of the aforementioned databases contain spontaneous data. However, even then the "spontaneity" of the recorded data remains questionable. Most of the spontaneous expressions were still recorded in unnatural/artificial settings (e.g., the University of Texas Database). Even if the subjects are shown movies that stimulate certain emotions, if they are aware of being filmed, their emotional response may not be *as* spontaneous anymore (e.g., will be masked or controlled). And even in the case where the subjects are unaware of being filmed, the laboratory situation may not encourage natural or usual emotion response.

Current bimodal/multimodal databases are yet to improve their features, content and annotation schemes to achieve the level of specifications listed above. Creating a spontaneous multimodal affect database is a challenging task involving ethical and privacy concerns together with technical difficulties (diverse and large set of subjects, high resolution, illumination, multiple sensors, consistency and repeatability within acceptable time limits). Given these restrictions, a database of directed emotional display has been the only alternative possible to date. Another challenging issue is that of creating a database that contains samples of both staged and spontaneous data in order to study the differences between these and how this procedure can be automated. No attempts are reported yet toward that goal.

10.3.3 Data Annotation

In order to foster development of affective multimodal systems, acquiring multimodal data is not enough; they need to be annotated and analyzed to form the ground truth for machine understanding of the human affective multimodal behavior.

As stated previously, when annotating or labeling affect data from face display, Ekman's theory of emotion universality [18] and the Facial Action Coding System (FACS), [17] are used. When it comes to annotating body gestures, unlike the AUs, there is not one common annotation scheme that can be adopted by all the research groups. The most common annotation has been *command-purpose annotation*, for instance calling the gesture as "rotate" or "click" gesture. Another type of annotation has been based on the *gesture phase*, e.g., "start of gesture stroke-peak of gesture stroke-end of gesture stroke". However, a more detailed annotation scheme, similar to that of FACS is needed. A general body gesture annotation scheme, possibly named the Body Action Unit Coding System (BACS), should include information and description as follows: *body part* (e.g., left hand), *direction* (e.g., up/down), *speed* (e.g., fast/slow), *shape* (hands made into fists), *space* (flexible/direct), *weight* (light/strong), *time* (sustained/quick), and *flow* (fluent/controlled) as defined by Laban and Ullman, [36, 37]. Additionally, temporal segments (neutral-start of gesture stroke-peak of gesture stroke-end of gesture stroke-neutral) of the gestures should be included as part of the annotation scheme.

Besides the aforementioned issues, affective state annotation in itself faces three main challenges: (a) the type of emotion encoded(e.g., some emotions are encoded easily compared to others), (b) the specific ability of the encoder (e.g., some cultures encode differently compared to others), and (c) specific, discriminative movement indicators for certain emotions versus indicators of the general intensity of the emotional experience (e.g., body movements are easily distinguishable from each other compared to face feature movements) [63]. Therefore, annotation of multimodal data is a very tiresome procedure overall as it requires extra effort and time to view and label the sequences with a consistent level of alertness and interest (e.g., it takes more than *one hour* to AU code *one minute* of face video). It is also not easy to obtain a significant number of qualified emotion coders representing various age and ethnic groups. Hence, obtaining the emotion- and quality-coding for all the data contained in multimodal databases is very difficult to achieve. Moreover, for the annotation purposes it is almost impossible to use emotion words that are agreed upon by everybody. The problem of what different emotion words are used to refer to the same emotion display is not, of course, a problem that is unique to this; it is by itself a topic of research for emotion theorists and psychologists. It is a problem deriving from the vagueness of language, especially with respect to terms that refer to psychological states [44].

As a rule of thumb, at least two main labeling schemes, in line with the psychological literature on descriptors of emotions, should be used: verbal categorical labeling (perceptually determined, e.g., happiness) and broad dimensional labeling: arousal (arousal–sleep) and valence (activated-Ũdeactivated). This labeling is in accordance with emotion theories in psychology: (a) Ekman's theory of emotion universality [18] and (b) Russell's theory of arousal and valence [55].

Taking into account these facts an ideal multimodal affect database should be annotated as follows: (a) Experimenters, preferably a group consisting of an expert in the affective computing field or an emotion researcher, should view and label the multimodal data. (b) Subjects' own evaluation should be obtained by asking the subjects after the recordings, to view and fill in a survey about their expressions. This feedback will form each subject's own evaluation of his affective state. (c) The multimodal data should additionally be annotated by independent human observers with different ethnic and/or cultural background in order to obtain independent interpretations. Moreover, it should be further analyzed whether being exposed to the expressions (hearing/seeing, etc.) from one sensor (face camera only) or another (body camera only), or from multiple sensors simultaneously (cameras and head-phones) affects the observer's interpretations. Annotation should also be analyzed statistically (e.g., how female observers annotate compared to male observers, how older observers annotate compared to younger observers, etc.).

10.3.4 Synchrony/Asynchrony Between Modalities

In affective multimodal systems, the kind of feature processing and fusion strategy to choose depends on the input data and modalities to be fused. There might be an inherent asynchrony between the modalities.

The asynchrony between modalities may be two-fold: (a) asynchrony in subject's signal production (e.g., the face movement might start earlier than the body move-ment) and (b) asynchrony in processing the signals coming from various sensing devices. For instance, assume that the computer is able to accept input coming from an affect sensitive mouse equipped with a psychological sensor and a camera cap-turing the body gestures. At the production level, the subject will continuously be sending psychological signals when touching the mouse whereas producing/making a gesture will probably take longer. At the processing level, the computer might complete processing data coming from the mouse sensor tens or even hundreds of milliseconds before the hand and/or body gestures are actually produced. Moreover, the difference between time responses of devices can be very large (a speech recog-nition system might need more time to recognize a word and link it to an affective state than a touch sensitive mouse to compute the current affective state of the user). If the modalities are not synchronized, then this might introduce pending recogni-tion results. Or the system might receive an information stream in an order which does not correspond to the real chronological order of user's actions. Incorrect fu-sion may occur due to the different time scales required to process data specified through distinct devices.

In the field of multimodal human-computer interaction (HCI) the researchers found that there exist synchrony issues based on the actual users. Experiments con-ducted on how humans integrate different modalities when working on a computer task showed that there are two different ways people integrate modalities: (a) syn-chronously and (b) asynchronously. Research shows that it is possible to group the users into either of these categories with simple experiments [45] (e.g., synchro-nous vs. asynchronous users). Interestingly, when the users were instructed in order to force them to change their integration pattern, the result was that the users would change the pattern for the time being but go back to their natural way of integration weeks later [45].

We are not yet aware how such findings will influence affective multimodal sys-tems. All of the existing affective multimodal systems have been developed without much knowledge about how the potential final users would combine the distinct modes to interact with the system. Depending on the context or task at hand, some users might tend to communicate with the affective system unimodally (e.g., an angry look only/ angry tone of voice only/ swearing words only, etc.), using one particular modality only. For another context or some other tasks, the users might tend to use multimodal channels (e.g., combining both their face and upper-body gestures for expressing their affective state). Therefore, prior to building an affective multimodal system, a detailed study of the tasks and users' multimodal behavior is needed to decide on how to customize the system according to the users' needs.

However, the main difference between the current human-computer interfaces and the foreseen affective multimodal systems is that affective systems will not need the user to adapt to them, instead they will adapt to the user, by mimicking HHI. Therefore, the user being synchronous or asynchronous might not be an issue at all.

10.3.5 Data Integration/Fusion

In affective computing, modality fusion is to combine and integrate, if possible, all incoming unimodal events into a single representation of the emotion most likely expressed by the user. Thus, fusion needs to synchronize the recognition and analysis components so that every unimodal event that could potentially contribute to the integrated meaning of a multimodal utterance is considered.

When it comes to integrating the multiple modalities the major issues are: (a) when to integrate the modalities (i.e., at what abstraction level to do the fusion) and (b) how to integrate the modalities (i.e., which criteria to use). Typically, the multimodal fusion problem is either done at the feature level in a maximum likelihood estimation manner or deferred to the decision level. To make the fusion issue tractable, the individual modalities are usually assumed independent of each other [12]. This simplification allows employing simple parametric models for the joint distributions that cannot capture the complex modalities' relationships. More importantly, this does not support mutual estimation (i.e., using the speech recognition information to inform the gestural recognition processing, or the processing of any other modality) [12].

Time has a primary role in fusion; a real-time multimodal interface needs to continuously attempt to combine input data. However, we cannot simply assume that input from all channels will be readily available at all times. Temporal analysis of affective multimodal data relies on time proximity [12, 43]: time-stamped features from different input channels are merged if they occur within a predefined time window. In a real-time multimodal system there might be such cases when only unimodal results are available. For instance, one of the recognizers finished recognition without any meaningful results and produced a time-out for reporting (e.g., the user is not using/touching the affect sensitive mouse). In this case the fusion component has to take only the hypotheses of the remaining modalities into account. However, in case all the modality recognizers reported some kind of error or time-out, the fusion component should inform the system about this and terminate the analysis procedure for the time window at hand [64].

Depending on how closely coupled the modalities are, there are three typical levels of integration: low level (data level), intermediate level (feature level/early fusion) and high level (decision level/semantic/late fusion).

Low-level fusion, also called data fusion, combines several sources of raw data to produce new raw data that is expected to be more informative and compact than the inputs (e.g., typically, in image processing, images presenting several spectral bands of the same scene are fused to produce a new image that ideally contains in a single channel all of the information available in the various spectral bands). Low-level fusion has not been particularly exploited in multimodal affective computing.

Feature-level fusion is performed by concatenating the feature vectors from each modality and using a single classifier, which uses the combined information to assign likelihoods to the recognizer's hypotheses. Early fusion enables use of some relationship between the different channels, for classification. To minimize the classification errors, some adaptation strategy can be adopted (e.g., weighting coefficients) [42]. When fusing the multimodal information at the feature level, the feature set can be quite large. Therefore, this level of fusion requires a large amount of data for the training and has high computational costs [12]. It is necessary to use a feature selection technique to find the features from both modalities that maximize the performance of the classifier(s). Typically, early integration architectures assume a strict time synchrony between the modalities.

In decision-level fusion each classifier processes its own data stream, and the two sets of outputs are combined at a later stage to produce the final hypothesis (sequential integration). Decision fusion (late integration) is most commonly found in CHI (e.g., speech and gesture) and is mostly applied to modalities that differ in the time characteristics of their features. Timing plays an important role and hence all fragments of the modalities involved are time-stamped and further integrated in conformity with some temporal neighborhood condition [12]. Designing optimal strategies for decision-level fusion has been of interest to researchers in the fields of pattern recognition, machine learning, and neural networks and more recently in data mining, knowledge discovery and data fusion. One approach, which has become popular across many disciplines, is based upon the combination of multiple classifiers, also referred to as an ensemble, committee or expert fusion. There has been some work on combining classifiers and providing theoretical justification for using simple operators such as majority vote, sum, product, maximum/minimum/median and adaptation of weights [21, 35]. Decision-level fusion can also be obtained in the following levels: (a) soft-level (a measure of confidence is associated with the decision) and (b) hard-level (the combining mechanism operates on single hypothesis decisions).

Which level of fusion to choose depends on the application. Usually, in automated affect analysis, late integration is chosen instead of early integration for the following reasons [12, 64]:

- The feature concatenation used in early integration results in a high-dimensional data space, making a large multimodal database necessary for robust statistical model training.
- Late integration allows asynchronous processing of the available modalities.

- Late integration provides greater flexibility in modeling. With late integration, it is possible to train different classifiers on different data sources and integrate them without retraining.
- Using late integration off-the-shelf recognizers can be utilized for single modalities (e.g., speech).
- Late integration allows adaptive channel weighting between the different modalities based on environmental conditions, such as the signal-to-noise ratio.

However, one should note that co-occurrence information is lost if late integration is chosen instead of early integration.

Wu and colleagues claim that fusion at the feature-level is appropriate for closely coupled and synchronized modalities (e.g., speech and lip-movements) [64]. They state that feature-level fusion tends to generalize if it consists of modes that differ substantially in the time scale characteristics of their features (e.g., speech and gesture input). Therefore, if the modalities are asynchronous but temporally correlated, like gesture and speech, decision-level integration is the most common way of integrating the modalities [64]. In summary, there is not a general consensus when fusing multiple modalities. Which fusion method to choose depends on the application, the modalities and users' integration patterns of these modalities (synchronous vs. asynchronous users). The right way to go could be by experimenting with each fusion technique separately and evaluating the performance of the system. Features could also be aggregated at different abstraction levels: lower visual primitives (e.g., optical flow) versus higher abstraction (e.g., face/body action units).

10.3.6 Information Complementarity/Redundancy

Various affective signals are typically congruent, and this redundancy facilitates efficient processing of multiple emotional signals in humans [59]. Indeed, people are able to accurately perceive multimodal signals in their daily lives. Moreover, although inputs from different modalities are processed in separate areas of the brain, our conscious experience is one of coherent, unified perceptions, reflecting that information becomes integrated across sensory modalities.

In multimodal systems, complementary input modalities provide the system with non-redundant information whereas redundant input modalities allow increasing both the accuracy of the fused information by reducing overall uncertainty and the reliability of the system in case of noisy information from a single modality [12]. In some cases the situation may be of mixed nature: for many emotions, multiple modalities may prove redundant; however, for some emotions, they might be the minimum informative set (i.e., indispensable). Complementary modalities need to be merged to result in the recognition of the best possible affective state. Then, what happens if the modalities contradict each other? For instance, in a bimodal system that consists of face and upper-body input, if the face input is recognized as displaying *happiness* but the body input is recognized as displaying *anger*, then what does/should the system output? More importantly, if feature-level fusion is applied, what happens with the features that are not supporting each other? What does the system output? None of the available multimodal systems have attempted to answer these questions. Most of the time the researchers assume that the modalities complement each other and increase the recognition accuracy. But what if there are cases or even emotion categories that are expressed in a non-complementary way?

In order to understand this problem we have to look at HHI and see how people use such information. Ambady and Rosenthal state that people look at the face and body more than any other channel when they judge nonverbal behavior [2]. However, they also note that exposing people to more than two channels causes ambiguity and people might get confused in their judgment (i.e., their recognition accuracy drops occasionally). Ekman found that the relative weight given to face expression, speech, and body cues depend both on the judgment task (i.e., what is rated and labeled) and the conditions in which the behavior occurred (i.e., how the subjects were simulated to produce the expression) [15]. Despite the aforementioned findings, there is no evidence in the actual HHI on how people attend to the various communicative channels (speech, face, body, etc.). Assuming that people judge these channels separately or the information conveyed by these channels is simply additive, is misleading [57]. It has been proposed that multimodal perception in humans occurs in three stages: evaluation, integration, and decision making [59]. First, each separable source of information is evaluated based on prototypes of particular emotional expressions. Next, integration involves the combination of the degree to which each source supports a given alternative (e.g., happy, angry, sad). Finally, a decision is made based on the amount of support for each alternative. When one source of information only weakly supports a possible alternative, other sources of information are given more influence [59]. Gunes and Piccardi found that in general, bimodal face-and-body data helps with resolving ambiguity carried by the face data alone. However, in some cases body adds ambiguity to the recognition [27]. What should be the strategy to follow in such cases? One option would be to leave out such data from the training stage of the multimodal affect recognizers hoping it will improve run-time accuracy. Another option would include training the system to ambiguous data and when the need arises forcing it to output results labeled as *ambiguous*. It would also be interesting to conduct some experiments on how humans operate in such cases and use those results while building an automated system. Another possible solution is introducing a simple mechanism such as *weight* or *confidence factor* for modeling uncertainty when ambiguous cases occur.

Redundant modalities instead would theoretically produce the same result either combined or taken separately. In this case, the system would need rules to identify the redundant input and possibly use some, if not all, of the redundant data. For instance, a user may utter swearing words while pointing his index finger at the computer, producing an angry face expression and touching the affect sensitive mouse (i.e., redundancy for the features coming from the same modality: gaze, head pose, face expression). If it is sufficient to use just the head pose, then it might save computational cost as estimating face expressions is much more complicated. In such cases, one of the user's actions should be ignored if not processed simultaneously. Redundancy check in such cases will help to avoid pending affect recognition results. However, this might work only for a particular combination of data as a real-time system needs to be designed for all possible scenarios.

If the number of extracted features from each modality are too many, redundancy might occur at the feature level. A possible way to deal with such redundant data would be applying some feature selection criteria prior to (for late fusion) or after fusion (for early fusion) by keeping in mind noise and non-ideal conditions.

10.3.7 Information Content of Modalities

In affective multimodal systems, which modality is more reliable than others? Input modes differ in both their information content and recognition accuracy. It is likely to be the case that we get good measures of some affective states but not others. In general, the reliability of the modalities depends on the task at hand, the ethnic background of the user and the emotion expressed (e.g., Japanese are taught to mask negative face expressions and to display emotions on a lesser scale compared to Westerners [7, 53]. Therefore, some modalities might be better than others for recognition of affective states in some cases.

Do people have modality preferences in the actual HHI? What does modality preference depend on? In general, there is no evidence of general modality dominance in humans; that is, people do not exhibit consistent preferences for either visual or auditory information [59]. However, there exist reports that modality preferences change with development and that environmental factors may influence modality preferences during middle childhood. The research reported in [59] suggests that the importance of auditory versus visual percepts is influenced by the meaning attached to particular emotions contained in each expression and by the familiarity of the individual expressing the emotion. For instance, children exhibited an auditory preference when presented with emotions expressed by their mothers, and a visual preference for emotions expressed by a stranger. This is consistent with the idea that vocal emotion may be more difficult to identify when expressed by unfamiliar individuals. Perceptual processing was also influenced by emotion: children demonstrated a preference for visual over auditory expressions of happiness. This finding is consistent with reports that although happiness is an easily recognizable face expression, it is more difficult to identify it in the voice. However, future research should examine whether people are able to deliberately control the deployment of their attention to one channel versus another or whether shifts in attention truly occur automatically.

Research suggests that humans are able to recognize an emotional expression in neutral-content speech with about 60% accuracy, and in face images with about 70–98% accuracy choosing from among about six different affective labels exhibited by actors [51]. Computer speech recognition that works at about 90% accuracy on neutrally-spoken speech (i.e., recognizing what is said) tends to drop to 50–60% accuracy on emotional speech (i.e., recognizing how it was said) [51]. Recent efforts indicate that combining audio and video signals for emotion recognition can give improved results. Computers have obtained 81% recognition accuracy on eight categories of emotion from physiological signals.

As can be seen from the aforementioned studies, overall recognition accuracy of different input modes cannot be assumed to be equally *reliable*. By *reliability* we mean the extent to which one modality yields the same recognition results as another modality. Some of the modalities might be corrupted by measurement noise and/or modeling errors. A single highly reliable modality alone may sometimes yield a correct decision, whereas its linear fusion with some other less reliable modality may give incorrect results. On other occasions, results obtained by fusion of two modalities may outperform those obtained from each modality alone. Even within the same mode, recognition accuracy varies considerably from one constituent to another [12]. A way of measuring the reliability of the modalities can potentially help improve the accuracy. As proposed in [19], for fusion of multiple modalities, certain reliability measures and rules can be created to compensate possible mis-classification errors of a certain classifier with other available classifiers and to end up with a more reliable overall decision.

10.4 Monomodal Systems Recognizing Affective Face or Body Movement

In this section we briefly review automatic systems that are capable of recognizing AUs, facial expressions/both, or affective body movements and analyze some of the recent representative works introduced during the period 2002–2006. For research on automatic face expression analysis up to year 2000, the reader is advised to see [20] and [48].

Existing automated face expression analyzers can be grouped in two categories, based on the type of face data they contain: (a) prototypical face expressions or (b) AU activations. The first group follows from [18] and contains face display of the six basic emotions (happiness, sadness, fear, disgust, surprise and anger) from either single images or image sequences. The second group of databases contain more subtle changes in face features (i.e., AUs) and are coded using the Facial Action Coding System (FACS) [17].

Herewith, we briefly mention representative systems introduced during the period 2002-2006. We provide a detailed comparison of the aforementioned systems against various criteria (feature extraction, recognition, and other criteria) in [24]. The reviewed works have been developed at the following universities: Delft University of Technology (TUDelft) [46], University of California, San Diego (UCSD) [4], Carnegie Mellon University (CMU) [10, 11], University of Amsterdam (UoA)/University of Illinois (UIUC) [9, 58], and Massachusetts Institute of Technology (MIT) ([34]). For simplicity, we use these names henceforth.

In general, the aforementioned systems are similar in the sense that they all (a) extract some features from the video sequences, (b) use the extracted features to feed a classifier/set of classifiers, and (c) produce an output either as detected AUs or emotions. They, however, differ in (a) the approaches they use for processing of the images, i.e., feature-based (detecting/tracking specific features such as the inner corners of the eyes) or region-based (measuring face motion in certain regions on the face such as mouth region), (b) features and number of features extracted, (c) outputting the recognition results as AUs or emotions, or both, and (d) the number of AUs detected.

According to the analysis provided in [24], the aforementioned face expression/AU recognizers show the following limitations:

- Some of them still require manual initialization of the features and/or face in the first frame of the face video.
- None of them is yet able to detect all 46 AUs present in FACS.
- Except for the TUDelft group, none of the groups has attempted to automate the annotation of the temporal segments of AUs.
- Except for the UIUC group, none of the groups has attempted to analyze sequences that contain multiple emotion displays.
- Except for the MIT system, all of the systems rely on the assumption that the first frame is a neutral frame.
- Although significant progress has been achieved, comparative evaluation of the various systems is still an issue, as they were all tested on different spontaneous behavior databases.

Most of the gesture-based systems exploited gesture input for command entry purposes [26] (e.g., selecting menus) using one-hand gestures only [57]. There exist gesture-based non-command interfaces (actions or events used to indirectly tune the system to the user's needs). Our focus is not on such systems. Instead, we briefly review automatic systems that attempt to recognize expressive body movement/gestures for affective computing.

In general, affective body recognition systems are similar to one another as they all do (a) motion segmentation, (b) object classification, (c) tracking, and (d) interpretation [57]. They differ in the detailed methodology they use for the aforementioned procedures. We mention here three representative systems that aim gesture/body motion analysis for affect/emotion recognition introduced in the literature in the context of CHI during the period 2002–2006. The reviewed works have been developed at the University of Genoa (UoG) by Camurri and his colleagues [62] and at the Rutgers University (RU) by Burgoon and her colleagues [6, 39]. We also provide a detailed comparison of these systems against various criteria (data, feature extraction and recognition) in [24].

The aforementioned body expression recognizers show the following limitations:

- The proposed approaches did not acquire data in natural settings over various periods of time.
- The proposed approaches have only trained their systems to automatically analyze 3 to 4 affective states.
- The proposed approaches have not attempted to analyze both the propositional (e.g., shape of the hand, posture of the head, posture of the body) and

non-propositional (e.g., speed, weight, trajectory, etc.) qualities of the expressive body display.
- The existing systems have not attempted automatic annotation of the temporal segments of the expressive body movements.
- The proposed approaches have not tested their system for culture dependency.

Overall, research in affective body expression recognition is relatively new and is clearly behind that of affective face analysis.

10.5 Multimodal Systems Recognizing Affect from Face and Body Movement

In this section we briefly review automatic systems that attempt to recognize affect from multimodal expressive behavior, face and body movement in particular. We present representative projects/systems introduced in the literature in the context of CHI (in chronological order) during the period 2002–2006. We also provide a detailed comparison of these systems against various criteria in Table 10.1 (data and feature extraction/tracking) and Table 10.2 (recognition).

10.5.1 Project 1: Multimodal Affect Analysis for Future Cars

Lisetti and her colleagues introduce the concept of utilizing a multimodal affective user interface for future cars in [38]. They argue that an ideal multimodal affective user interface should at least integrate the visual, haptic, and auditory modalities. The proposed system is also intended to receive input from linguistic tools in the form of linguistic terms for emotion concepts. However, of all the modalities proposed initially for the multimodal system, the authors seem to have integrated the face expression and kinesthetic part only. Although they claim that their system can perform real-time face expression recognition, the details of the system are not presented.

In order to map certain physiological signals to certain emotions, the authors designed an experiment. Ten undergraduate and graduate college students (5 female, 5 male) participated in their 35 minute experiment. During the experiment they elicited five emotions (neutral, anger, fear, sadness, and frustration) and measured three physiological signals (galvanic skin response (GSR), heartbeat, and temperature). For their experiment, they designed a slide show, which they presented to the participants. The slide show started with a relaxation period followed by a picture/movie clip/scenario presented to the participant in order to elicit one of the five emotions. This was followed by another relaxation period and another emotion elicitation period. This process was repeated until all the emotions were elicited. They used two different algorithms to analyze the data collected: k-Nearest Neighbor

Table 10.1 A comparison of existing multimodal systems analyzing expressive face and/or body movement: data and feature extraction.

criteria	Lisetti et al.[38]	Balomenos et al.[3]	Kapoor&Picard[33]	Gunes&Piccardi[28]
affect sensing	face expression and physiological signals(galvanic skin response (GSR), skin temperature, heartbeat)	upper-body movement combined with face exp.	a camera and a pressure sensing chair	two cameras, one for upper body and one for face
sensors used	1 camera and 1 BodyMedia Sense wear for physiological signals	1 camera	chair sensor and 1 camera	2 cameras
natural settings/spontaneous data?	no	no	yes	no
context used	future cars	no	game context	HHI/CHI
methods used for recordings	Body Media Sense wear for recording the physiological signals. Participants were shown cuts from movies while obtaining the recordings.	not specified for the face; body gesture sequences were obtained of 3 males, with maximum duration of 3 sec, captured by a web-camera at a rate of 10 fps.	Children were asked to play a game for about 20 minutes, 3 channels of data (face, posture and game information) were recorded.	participants were asked to display face and body gestures according to the vignettes provided for each emotion category.
# of subjects	10	3	8 children	4
# of data instances	from 10 subjects 10 instances of 35 min	not specified for face. 90 sequences for the body gestures	61 samples of high interest, 59 samples of low interest and 16 samples of taking a break	54 videos,27 for face and 27 for body
affective states	5:neutral, anger, fear, frustration, sadness	6:joy, sadness, anger, disgust, fear, surprise	3:low interest, high interest, refreshing	6:happiness, disgust, fear, anger, uncertainty, anxiety
data labeled by independent observers?	no	no	yes, several teachers for the face modality	no
quality of movement?	no	no	no	no
usage of space?	no	no	no	yes, as part of the feature set
upper body?	no	yes	yes	yes
body posture?	no	no	yes	no
hand movement/posture?	no	yes	no	yes
relationship btw. body parts?	no	no	no	yes
video/ static image analysis?	unknown	static frame for face, video based for body	both static frame-by-frame analysis and sequence-based analysis	feature extraction and tracking was done for a whole video
features used	physiological signals, number not specified	extracting the eyes and the lips, tracking the face points (lips, eyes, eyebrows) from MPEG-4 compatible animation; moving skin masks for hands and head	Upper Face(brow shape, eye shape, likelihood of nod/shake/blink); Lower Face (probability of fidget/smile); current posture and level of activity; level of difficulty, state of the game	more than 100 features for each modality, used feature selection to minimize the number
feature extraction methods used	body media wear's analysis tools	not explained for the face, assuming head and hands are initially located at certain places in the frame; tracking the centroid of the head and the hands with a skin mask	pupils tracked by infrared camera; the pupil positions used for head-nod/head-shake detection; the mouth localized by extracting two real numbers; smiles and fidgets; postures are recognized using two matrices of pressure sensors.	for face: geometric based methods, skin color segmentation,histogram equalization, thresholding, edge maps,X- and Y- axis projection of the face histogram, min-max analysis
tracking of features?	NA	yes, hands and head; not specified for face	yes	no for face; yes for upper body(hand and face region)

Table 10.2 A comparison of existing multimodal systems analyzing expressive face and/or body movement: recognition.

criteria	Lisetti et al.[38]	Balomenos et al.[3]	Kapoor&Picard[33]	Gunes&Piccardi[28]
fully automatic?	no	no, cropping the face region manually	yes	no
feature selection?	no	no	yes	yes, best-first search method
recognition method	k-Nearest Neighbor and discriminant functions	HMMs for gesture	SVMs, HMMs, mixture of Gaussians	BayesNet classifier in WEKA
training data	no training applied	60 videos from 3 people for body gestures	50% of the data (68 videos)	50% of the data (27 videos)
testing data	method tested on all the data obtained	30 videos from 3 people for body gestures	50% of the data (68 videos)	50% of the data (27 videos)
data ethnically diverse?	unknown	no	not specified	yes
fusion method	data not fused	decision level, predefined weights	feature- and decision-level fusion	feature- and decision-level fusion
recognition accuracy for face modality	not reported	85%	upper face 67%, lower face 53%	75%
recognition accuracy for body modality	NA	94%	posture 82%	91%
other modality	NA	NA	game status 57%	NA
multimodal recognition accuracy	kNN:72%,DFA: 74%,MBG: 84% for 6 emotions	not provided	86%	feature level:94%; decision level:sum 91%, product 87%, weight 80%
handles missing channels/noisy labels?	no	no	yes	no
automatic recognition compared to human recognition?	no	no	yes, for face	no
culture dependency tested?	no	no	no	no
automatic temporal segment annotation?	no	no	no	no

Algorithm and Linear Discriminant functions. The best results were provided by the k-Nearest Neighbor Algorithm with the following recognition accuracies: neutral (100%), anger (100%), fear(80%), frustration (80%) and sadness (60%).

Overall, they implemented separately two portions of the system they proposed: face expression recognition and physiological signal analysis. However, they did not attempt to acquire bimodal data using the camera and the physiological sensor simultaneously. They also did not report on data fusion and compare the monomodal and bimodal recognition results.

10.5.2 Project 2: Emotion Analysis in Man-Machine Interaction Systems

Balomenos and his colleagues combined face expressions and hand gestures for recognition of prototypical emotions by using face points from MPEG-4 compatible animation and defining certain hand movements under each emotion category [3]. They recognize six emotion categories, namely: anger, fear, disgust, joy, sadness and surprise.

Face detection is performed through detection of skin segments or blobs, merging them based on the probability of their belonging to a face area, and identification of the most salient skin color blob or segment. Primary face features, such as eyes, mouth and nose, are dealt with by major discontinuities on the segmented, arbitrarily rotated face. Following face detection, morphological operations are used to define first the most probable blobs within the face area to include the eyes and the mouth. Searching through gradient filters over the eyes and between the eyes and mouth provide estimates of the eyebrow and nose positions. Based on the detected face feature positions, feature points are computed and evaluated. They achieved 85% accuracy for emotion recognition from face features alone. The system is further based on apriori knowledge that the head is expected to be located in the middle area of the upper half of the frame and the hand segments near the respective lower corners. They track the position of the centroid of the head and the hands over time. They experimented on gesture sequences of three male subjects, with maximum duration of three seconds, that were captured by a typical web-camera at a rate of 10 frames per second. For each of the gesture classes 15 sequences were acquired, 3 used for initialization of the HMM model, 7 for training and parameter re-estimation and 5 for testing. Each training sequence consisted of 15 frames (selected manually and off-line). Testing sequences were subsampled at a rate of 5 frames per second. An overall recognition rate of 94% was achieved for emotion recognition from hand gestures.

They fused the results from the two subsystems at a decision level using pre-defined weights (0.75 for face modality and 0.25 for body modality). However, in [3] they do not report the recognition accuracy for the fused data. It is also not clearly explained in their paper how the recordings were obtained or how experiments were conducted for the bimodal data.

10.5.3 Project 3: Multimodal Affect Recognition in Learning Environments

Kapoor and Picard described a project on machine recognition of affect using multiple modalities [33]. They looked at the problem of detecting the affective states of high interest, low interest, and a state called "taking a break" (a forward-backward postural fidget/stretching) in a child who is solving a puzzle. To this aim, they combined sensory information from the face, the postures and the state of the puzzle using a unified Bayesian approach based on a mixture of Gaussian Process (GP) classifiers.

Overall, the proposed system extracts the following features for each channel: upper face (eyebrow shape, eye shape, likelihood of nod, likelihood of shake, likelihood of blink); lower face (probability of fidget, probability of smile); posture (current posture, level of activity); and game (level of difficulty, state of the game).

Postures were recognized using two matrices of pressure sensors placed on a chair. An in-house built version of the IBM Blue Eyes Camera was used to track pupils using two sets of IR LEDs. The rest of the face feature extraction techniques were based on the pupil detection and tracking module. Tracked pupils were first used to recover shape information of eyes and the eyebrows. The details of the automatic upper face feature extraction and AU detection can be found in [34]. Second, the pupil positions were passed to an HMM-based head-nod and head-shake detection system, which provided the likelihoods of head-nods and head-shakes. Another HMM used the radii of the visible pupil as inputs to produce the likelihoods of blinks. Third, the detected pupil positions were used to localize the candidate mouth region. Two real numbers corresponding to two kinds of mouth activities were extracted: smiles and fidgets. To this aim, the sum of the absolute difference of pixels of the extracted mouth image in the current frame with the mouth images in the last 10 frames were utilized. The assumption used was that a large difference in images should correspond to mouth movements (i.e., the fidgets). The probability of a smile was computed by a previously trained support vector machine (SVM).

The idea behind GP classification is that the hard labels depend upon hidden soft-labels, which are assumed to be jointly Gaussian with the covariance between outputs specified using a kernel function applied to inputs. The task then becomes inferring the label for an unlabeled data point. See [33] for details. The inferral of the class probability of an unlabeled data was obtained by using a mixture of GPs. All the posterior probabilities obtained from all different classifiers for each sensor were used for the final decision.

The final database used to train and test the system included 8 different children with 61 samples of high interest, 59 samples of low interest and 16 samples of taking a break. The experiments were performed to classify the state of interest (65 samples) vs. uninterest (71 samples). The experimental methodology was to use 50% of the data for training and use the rest for testing. The classification results obtained by GPs for each individual modality are as follows: upper face 67%, lower face 53%, posture 82%, and game 57%. The fusion was obtained by a unified Bayesian approach based on a mixture of GP classifiers where classification using each channel is learned via Expectation Propagation. This resulted in 87% accuracy. The posture channel seemed to classify the modalities best, followed by features from the upper face, the game and the lower face. Fusion significantly outperformed classification using the individual modalities [33]. The Mixture of GPs also outperformed mixed rule-based decision fusion of the individual SVM and GP classifiers. Further to the comparison with the individual modalities, comparison of the mixture of GPs with an HMM-based expert-critic scheme and a naive feature-level fusion were also provided. They tested a naive feature level fusion where they used -1 as a value of all those observations that were missing, thus fusing all the channels into one single vector. The Mixture of GPs performed better than other approaches both in the case of data where all channels were present and incomplete data.

Note that Kapoor and Picard did not test their system for unseen subjects; accuracy of their system might be lower for totally unseen subjects.

10.5.4 Project 4: FABO-Fusing Face and Body Gestures for Bimodal Emotion Recognition

Gunes and Piccardi created a bimodal database that consisted of recordings of face expressions alone and combined face and body expressions [25]. They recorded the sequences simultaneously using two fixed cameras with a simple setup and uniform background. The FABO database has already been used for the validation of the approach proposed in [28] which could not have been possible with any existing databases due to their lack of combined affective face and body displays. In [28], the authors presented an approach to automatic visual recognition of expressive face and upper-body gestures from video sequences suitable for use in a vision-based affective multimodal framework. The feature vectors consisted of displacement measures between two major frames; namely, a frame with the neutral expression ("neutral frame") and one where the expression is at its apex ("expressive frame"). The following steps were taken for extraction of face features: (a) skin color segmentation based on HSV color space was applied, (b) the face region was obtained by choosing the largest connected component among the candidate skin areas, (c) closing (dilation and erosion) was employed and the contour of the face was obtained together with the filled face region. For feature extraction two basic methods were applied: (a) the gray-level information of the face region combined with edge maps and (b) the min-max analysis to detect the eyebrows, eyes, mouth and chin, by evaluating the topographic gray-level relief. After detecting the key features in the neutral frame and defining the bounding rectangles for face features, the temporal information in subsequent frames was considered by computing the optical flow in such bounding rectangles. The wrinkle changes were analyzed by using edge density per unit area against a threshold.

The following steps were taken for body feature extraction: (a) in each frame a segmentation process based on a background subtraction method was applied in order to obtain the silhouette of the upper body, (b) thresholding followed by noise cleaning and morphological filtering was applied, (c) a binary connected component operator was used to find the foreground regions, and small regions were eliminated. A set of features were generated for the detected foreground object, including its centroid, area, bounding box and expansion/contraction ratio for comparison purpose. For the segmentation and tracking of the body parts, the face and the hands were located exploiting skin color information. The centroid of these regions were calculated in order to use them as reference points for the body movement. The Camshift technique was employed for tracking the hands and comparison of bounding rectangles was used to predict their locations in subsequent frames.

For the experiments, they processed 54 sequences in total, 27 for face and 27 for body from four subjects, by using only the "neutral" and "expressive" or "apex" frames for training and testing. Half of these were used for training and the other half for testing purposes. For monomodal emotion recognition, the best recognition results were obtained with the BayesNet classification algorithm (76% for face and 90% for body). For bimodal emotion recognition both feature-level and decision-level fusion were performed. For feature level fusion, a feature selection method was utilized prior to classification. On a dataset consisting of 412 training and 386 testing instances, with 14 attributes, BayesNet provided the best classification accuracy (94% recognition accuracy). For decision-level fusion the sum, product, and weight rules (0.7 for the face modality and 0.3 for the body modality) were used. The late fusion results obtained were as follows: 91% recognition accuracy for sum, 87% product, and 80% for weight criteria.

Overall, the system of Gunes and Piccardi has the following limitations: (a) as the neutral and apex frames need to be chosen manually, the system remains semi-automatic; (b) although they perform tracking they do not utilize full-length expressive video sequences; (c) their experiment tests unseen instances from the same subjects used for the training phase, but they did not test their system for unseen subjects, and accuracy of their system might be lower for totally unseen subjects.

10.6 Multimodal Affect Systems: The Future

This chapter focused on affective multimodal systems taking face and body modalities as input, as these systems have been introduced in the last few years and the interest is relatively new. Representative systems were described and compared.

One major finding of the survey on the multimodal affect systems is that body gestures or postures provided better information than other modalities for affect recognition [3, 28, 33]. Although it was previously stated that analyzing both the propositional (e.g., thumbs up) and non-propositional gestures (e.g., how smooth/jerky the movement is) might be more promising [30], none of the existing systems have attempted this. In summary, the aforementioned multimodal systems have the following limitations:

- The existing systems have not attempted automatic annotation of the temporal segments of affect modalities.
- The existing systems have not acquired natural data over various periods of time (in different sessions).
- The existing systems have not analyzed the quality of body movement (i.e., how fast/smooth/jerky movements are).
- The existing systems have not captured and analyzed whole body movement (i.e., mostly upper-body movement was analyzed).
- The existing systems have not attempted to analyze the coordination of the input modalities (i.e., what happens with the face when hands start moving, etc.).
- All of the proposed systems can handle limited number of modalities (2 or 3), none of them have combined or explored all possible modalities for automatic affect recognition: speech (i.e., linguistic terms/words), audio (i.e., pitch, etc.), face expression/AUs, body posture, expressive body gesture, physiological sensing, brain signals, olfactory signals, etc.

In order to understand where the current state-of-the art in multimodal affect recognition stands compared to what is aimed, we now discuss the features of an ideal multimodal affective system. An ideal multimodal affect analyzer should have the following features:

- Achieve automatic real-time multimodal data acquisition, processing and affective state recognition.
- Handle and recognize all possible affective states: expressed synchronously or asynchronously, expressed with intention (e.g., joy) or without intention (e.g., boredom, fatigue).
- Handle large head or body movements as well as moving subjects in various environments (e.g., office or house, not just restricted to one chair or room).
- Deal with both posed and spontaneous data where the subject *is not aware of the recording*, hence will not restrain himself/herself unlike the case when s(he) is part of an experiment, will express emotions due to real-life event or trigger of events (e.g., stressed at work).
- Handle occurrences of *occlusions* (e.g., hands occluding each other or hand occluding the face), noise (e.g., in audio recordings) and missing data.
- Obtain and analyze input from *multiple sensing devices* (e.g., multiple cameras & microphones & haptic/olfactory/taste/brain sensors, etc.).
- Handle *non-uniform and noisy (lighting/voice recording) conditions*.
- Handle *long sessions*, expanding between one day and possibly a couple of weeks, capturing *all variations* of expressive1- expressive2- expressive3- neutral behavior in every possible order or combination.
- Deal with subjects of *diverse age, gender and ethnic background*.
- *Adaptive* to user, task and context.

Every research group agrees that multiple modalities should be explored in order to understand which channels provide better information for automatic affect/emotion recognition. Looking at Figure 10.1, in general, among the available modalities in HHI, for affective CHI, sight, sound and physiological sensing have been explored to some extent. Although there is a recent interest in the *thought* modality (i.e., brain–computer interfaces) [52], channels such as *smell* and *taste* remain totally unexplored. Overall, multimodal affect systems are still in their infancy. Further progress is mandatory in order to achieve natural multimodal affective CHI comparable to that of HHI.

References

1. *The smartkom corpora:* *http://www.phonetik.uni–muenchen.de/ bas/ basmulti-modaleng.huml#smartkom* (Access date: 27 November, 2006).
2. N. Ambady and R. Rosenthal, *Thin slices of expressive behavior as predictors of interpersonal consequences: A meta–analysis*, Psychological Bulletin **11** (1992), no. 2, 256–274.
3. T. Balomenos, A. Raouzaiou, S. Ioannou, A. Drosopoulos, and K. Karpouzis, *Emotion analysis in man–machine interaction systems*, Proc. of the Workshop on Multimodal Interaction and Related Machine Learning Algorithms, 2004, pp. 318–328.
4. M.S. Bartlett, G. Littlewort, M. Frank, C. Lainscsek, I. Fasel, and J. Movellan, *Fully automatic facial action recognition in spontaneous behavior*, Proc. of the IEEE Int. Conf. on Automatic Face and Gesture Recognition, 2006, pp. 223–230.
5. B. Braathen, M.S. Bartlett, G. Littlewort-Ford, E. Smith, and J.R. Movellan, *An approach to automatic recognition of spontaneous facial actions*, Proc. of the Int. Conf. on Automatic Face and Gesture Recognition, 2002, pp. 231–235.
6. J. K. Burgoon, M. L. Jensen, T. O. Meservy, J. Kruse, and J. F. Nunamaker, *Augmenting human identification of emotional states in video*, Proc. of the Int. Conf. on Intelligent Data Analysis, 2005.
7. J.K. Burgoon, D.B. Buller, and G.W. Woodall, *Nonverbal communication: The unspoken dialogue*, Harper and Row, New York, 1989.
8. C.Y. Chen, Y.K. Huang, and P. Cook, *Visual/acoustic emotion recognition*, Proc. of the IEEE Int. Conf. on Multimedia and Expo, 2005, pp. 1468–1471.
9. I. Cohen, N. Sebe, A. Garg, L. Chen, and T.S. Huang, *Facial expression recognition from video sequences: temporal and static modeling*, Computer Vision and Image Understanding **91** (2003), 160–187.
10. J.F. Cohn, L.I. Reed, Z. Ambadar, Jing X., and T. Moriyama, *Automatic analysis and recognition of brow actions and head motion in spontaneous facial behavior*, Proc. of the IEEE Int. Conf. on Systems, Man and Cybernetics, vol. 1, 2004, pp. 610–616.
11. J.F. Cohn, L.I. Reed, T. Moriyama, Jing X., K. Schmidt, and Z. Ambadar, *Multimodal coordination of facial action, head rotation, and eye motion during spontaneous smiles*, Proc. of the IEEE Int. Conf. on Automatic Face and Gesture Recognition, 2004, pp. 129–135.
12. A. Corradini, M. Mehta, N.O. Bernsen, and J.-C. Martin, *Multimodal input fusion in human computer interaction on the example of the on–going nice project*, Proc. of the NATO–Asi Conf. on Data Fusion for Situation Monitoring, Incident Detection, Alert and Response Management, 2003, pp. 223–234.
13. M. Coulson, *Attributing emotion to static body postures: Recognition accuracy, confusions, and viewpoint dependence*, Nonverbal Behavior **28** (2004), no. 2, 117–139.
14. M. DeMeijer, *The contribution of general features of body movement to the attribution of emotions*, Journal of Nonverbal Behavior **13** (1989), no. 4, 247–268.
15. P. Ekman, *Emotions in the human faces*, 2 ed., Studies in Emotion and Social Interaction, Cambridge University Press, 1982.
16. P. Ekman and W. V. Friesen, *Nonverbal behavior in psychotherapy research*, Research in Psychotherapy (1968), 179–216.
17. P. Ekman and W. V. Friesen, *The facial action coding system: A technique for measurement of facial movement*, Consulting Psychologists Press, San Francisco, CA, 1978.

18. P. Ekman and W.V. Friesen, *Unmasking the face: A guide to recognizing emotions from facial clues*, Prentice Hall, Englewood Cliffs, NJ, 1975.

19. E. Erzin, Y. Yemez, and A. M. Tekalp, *A theoretical and experimental analysis of linear combiners for multiple classifier systems*, IEEE Trans. on Multimedia **7** (2005), 840–852.

20. B. Fasel and J. Luettin, *Automatic facial expression analysis: a survey*, Pattern Recognition **36** (2003), 259–275.

21. G. Fumera and F. Roli, *A theoretical and experimental analysis of linear combiners for multiple classifier systems*, IEEE Trans. on Pattern Analysis and Machine Intelligence **27** (2005), 942–956.

22. H.J. Go, K.Ch. Kwak, D.J. Lee, and M.G. Chun, *Emotion recognition from the facial image and speech signal*, Proc. of the SICE Annual Conf., vol. 3, 2003, pp. 2890–2895.

23. D. Goleman, *Emotional intelligence: why it can matter more than IQ*, Bantam Books, New York, 1995.

24. H. Gunes, *Vision-based multimodal analysis of affective face and upper-body behaviour*, University of Technology, Sydney, Australia, 2007, Ph.D. Dissertation.

25. H. Gunes and M. Piccardi, *A bimodal face and body gesture database for automatic analysis of human nonverbal affective behavior*, Proc. of the Int. Conf. on Pattern Recognition, vol. 1, 2006, pp. 1148–1153.

26. H. Gunes and M. Piccardi, *Creating and annotating affect databases from face and body display: A contemporary survey*, Proc. of the IEEE Int. Conf. on Systems, Man and Cybernetics, 2006, pp. 2426–2433.

27. H. Gunes and M. Piccardi, *Observer annotation of affective display and evaluation of expressivity: Face vs. face-and-body*, Proc. of the HCSNet Workshop on the Use of Vision in Human-Computer Interaction, 2006, pp. 35–42.

28. H. Gunes and M. Piccardi, *Bi–modal emotion recognition from expressive face and body gestures*, Journal of Network and Computer Applications **30** (2007), no. 4, 1334–1345.

29. N. Hadjikhani and B. De Gelder, *Seeing fearful body expressions activates the fusiform cortex and amygdala*, Current Biology **13** (2003), 2201–2205.

30. E. Hudlicka, *To feel or not to feel: the role of affect in human–computer interaction*, Int. Journal of Human–Computer Studies **59** (2003), no. 1–2, 1–32.

31. A.M. Isen, *Positive affect and decision making*, Handbook of Emotions (M. Lewis and J. Haviland, eds.), Guilford, New York, 2000.

32. T. Kanade, J.F. Cohn, and Y.L. Tian, *Comprehensive database for facial expression analysis*, Proc. of the IEEE Int. Conf. on Automaitc Face and Gesture Recognition, 2000, pp. 46–53.

33. A. Kapoor and R. W. Picard, *Multimodal affect recognition in learning environments*, Proc. of the ACM Int. Conf. on Multimedia, 2005, pp. 677–682.

34. A. Kapoor, Y. Qi, and R.W. Picard, *Fully automatic upper facial action recognition*, Proc. of the IEEE Int. Workshop on Analysis and Modeling of Faces and Gestures, 2003, pp. 195–202.

35. J. Kittler, M. Hatef, R.P.W. Duin, and J. Matas, *On combining classifiers*, IEEE Trans. on Pattern Analysis and Machine Intelligence **20** (1998), no. 3, 226–239.

36. R. Laban and F.C. Lawrence, *Effort*, 2 ed., MacDonald and Evans, London, 1974.

37. R. Laban and L. Ullmann, *The mastery of movement*, 4th revision ed., Princeton Book Company Publishers, Princeton, NJ, 1988.

38. C. L. Lisetti and F. Nasoz, *Maui: A multimodal affective user interface*, Proc. of the ACM Int. Conf. on Multimedia, 2002, pp. 161–170.

39. S. Lu, G. Tsechpenakis, D.N. Metaxas, M.L. Jensen, and J. Kruse, *Blob analysis of the head and hands: A method for deception detection*, Proc. of the Annual Hawaii Int. Conf. on System Science, 2005, pp. 20–29.

40. S. Mader, C. Peter, R. Goecke, R. Schultz, J. Voskamp, and B. Urban, *A freely configurable, multi–modal sensor system for affective computing*, Proc. of Affective Dialogue Systems: Tutorial and Research Workshop, 2004, pp. 313–318.

41. A. Mehrabian, *Communication without words*, Psychology Today, vol. 2, 1968.

42. S. Nakamura, *Statistical multimodal integration for audio–visual speech processing*, IEEE Trans. oon Neural Networks **13** (2002), no. 4, 854–866.

43. L. Nigay and J. Coutaz, *A generic platform for addressing the multimodal challenge*, Proc. of the Conf. on Human Factors in Computing Systems (CHI), 1995.

44. A. Ortony and T. J. Turner, *What's basic about basic emotions?*, Psychological Review **97** (1990), 315–331.

45. S. Oviatt, R. Coulston, S. Tomko, B. Xiao, R. Lunsford, M. Wesson, and L. Carmichael, *Toward a theory of organized multimodal integration patterns during human–computer interaction*, Proc. of the Int. Conf. on Multimodal Interfaces, 2003, pp. 44–51.

46. M. Pantic and I. Patras, *Dynamics of facial expression: Recognition of facial actions and their temporal segments from face profile image sequences*, IEEE Trans. on Systems, Man and Cybernetics, Part B **36** (2006), no. 2, 433–449.

47. M. Pantic and L. Rothkrantz, *Toward an affect sensitive multimodal human–computer interaction*, Proc. of the IEEE **91** (2003), no. 9, 1370–1390.

48. M. Pantic and L. J. M. Rothkrantz, *Automatic analysis of facial expressions: The state of the art*, IEEE Trans. on Pattern Analysis and Machine Intelligence **22** (2000), no. 12, 1424–1445.

49. R. W. Picard, *Affective computing: challenges*, Int. Journal of Human–Computer Studies **59** (2003), no. 1–2, 55–64.

50. R.W. Picard, *Affective computing*, MIT Press, Cambridge, MA, 1997.

51. R.W. Picard, E. Vyzas, and J. Healey, *Toward machine emotional intelligence: analysis of affective physiological state*, IEEE Trans. on Pattern Analysis and Machine Intelligence **23** (2001), no. 10, 1175–1191.

52. T. Pun, T.I. Alecu, G. Chanel, J. Kronegg, and S. Voloshynovskiy, *Brain–computer Interaction Research at the Computer Vision and Multimedia Laboratory, University of Geneva*, IEEE Trans. on Neural Systems and Rehabilitation Engineering **14** (2006), 210–213.

53. S. J. Ramsey and J. Birk, *Training North Americans for Interaction with Japanese: Considerations of language and communication style*, The Handbook of Intercultural Training (D. Landis and R. W. Brislin, eds.), Area Studies in Intercultural Training, vol. 111, Pergamon Press, New York, 1983.

54. B. Reeves and C. Nass, *The media equation: How people treat computers, television and new media like real people and places*, Cambridge University Press, London, 1996.

55. J. A. Russell, *A circumplex model of affect*, Journal of Personality and Social Psychology **39** (1980), 1161–1178.

56. P. Salovey and J.D. Mayer, *Emotional intelligence*, Imagination, Cognition, and Personality **9** (1990), 185–211.

57. N. Sebe, I. Cohen, and T.S. Huang, *Multimodal emotion recognition*, Handbook of Pattern Recognition and Computer Vision, World Scientific, 2005.

58. N. Sebe, M. S. Lew, I. Cohen, Y. Sun, T. Gevers, and T. S. Huang, *Authentic facial expression analysis*, Proc. of the IEEE Int. Conf. on Automatic Face and Gesture Recognition, 2004, pp. 517–522.

59. J. E. Shackman and S. D. Pollak, *Experiential influences on multimodal perception of emotion*, Child Development **76** (2005), 1116–1126.

60. M. Song, J. Bu, C. Chen, and N. Li, *Audio–visual based emotion recognition– a new approach*, Proc. of the IEEE Conf. on Computer Vision and Pattern Recognition, vol. 2, 2004, pp. 1020–1025.

61. A.J. Toole, J. Harms, S.L. Snow, D.R. Hurst, M.R. Pappas, J.H. Ayyad, and H. Abdi, *A video database of moving faces and people*, IEEE Trans. on Pattern Analysis and Machine Intelligence **27** (2005), no. 5, 812–816.

62. G. Volpe, *Computational models of expressive gesture in multimedia systems*, Faculty of Engineering, University of Genova, Genova, 2003, Ph.D. Dissertation.

63. H.G. Wallbott, *Bodily expression of emotion*, European Journal of Social Psychology **28** (1998), 879–896.

64. L. Wu, S.L. Oviatt, and P.R. Cohen, *Multimodal integration–a statistical view*, IEEE Trans. on Multimedia **1** (1999), no. 4, 334–341.

Chapter 11
Importance of Vision in Human-Robot Communication: Understanding Speech Using Robot Vision and Demonstrating Proper Actions to Human Vision

Yoshinori Kuno, Michie Kawashima, Keiichi Yamazaki, and Akiko Yamazaki

Abstract Vision plays an important role in communication. We observe situations and human actions through vision to obtain information necessary for smooth communication. In order to develop robots that can coexist with humans, it is necessary to take into account vision. There are two crucial tasks robots have to be able to perform. First, they have to be able to obtain visual information from human action. Second, they have to be able to respond (i.e., move their bodies) in such a way that their actions can convey proper information to human vision. In this paper, we will report on the results of our research in relation to these two points. In relation to the first, we will report on a helper robot that can respond to simplified utterances with deixis or ellipsis by recognizing human actions with vision. In relation to the second, we will discuss a museum guide robot that can move its head in a communicative way while explaining exhibits to visitors.

11.1 Introduction

Vision plays an important role in communication in various ways. We observe situations and human actions through vision and obtain information necessary for communication to proceed smoothly. In order to develop robots that can coexist with humans in various capacities (e.g., museum guides, elderly care), it is necessary to

Yoshinori Kuno
Graduate School of Science and Engineering, Saitama University, Saitama, Japan,
e-mail: kuno@cv.ics.saitama-u.ac.jp

Michie Kawashima, Keiichi Yamazaki
Faculty of Liberal Arts, Saitama University, Japan,
e-mail: yamakei@post.saitama-u.ac.jp, kawashima411@nifty.com

Akiko Yamazaki
School of Systems Information Science, Future University-Hakodate, Japan,
e-mail: akikoy@fun.ac.jp

D. Monekosso et al. (eds.), *Intelligent Environments*, Advanced Information and Knowledge Processing, DOI: 10.1007/978-1-84800-346-0_11,
© Springer-Verlag London Limited 2009

consider how robots can manage communication through vision. There are two crucial tasks robots need to be able to do for this purpose. First, they have to be able to obtain visual information from human action. Second, they have to be able to move their bodies in such a way that their actions can convey proper information to human vision. This paper presents the results of our research on such vision-and-action issues in human-robot communication.

Speech is an important means of the human interface for helper robots, which are becoming increasingly useful in a rapidly aging society. Thus, studies have investigated robots with speech interface [1, 2]. In addition to explicit utterances (e.g., Get the red book), robots must also be able to deal with implicit utterances such as those that contain deixis (e.g., Get that for me) and ellipsis (e.g., Get for me), as these are common phenomena in daily conversation. Their use in relation to context and a priori knowledge has been investigated in the fields of natural language and speech understanding [3, 4]. We may omit or mention things that are apparent to the listener(s) in the immediate scene. For example, we may say, "Get that for me," even though the object referenced by 'that' was not mentioned in the prior discourse. When the object is available in the immediate scene and the listener appears to be gazing in its direction, we assume that he or she can identify the object referenced by 'that'. In order to be "user-friendly", robots should be able to respond to utterances based on information supplied in both talk and the visual field. Grice has proposed several conversational maxims [5]. According to one, conversation is a cooperative endeavor between speaker and listener where both offer necessary and sufficient related information briefly and clearly. Based on this, we assume we can obtain important information in part through vision about things not mentioned in speech. Actually, there are various inexplicit utterances not interpretable through vision alone (e.g., prior shared experience between speaker and hearer). These types of utterances are beyond the scope of the present endeavor.

In this paper, we present a method for understanding speech by using computer vision. In particular, we are developing a helper robot that is able to bring an object to the speaker that the speaker requests through speech [6]. Here, we show how the robot can also attend to requests in relation to the visual field. This is an initial attempt to use computer vision to properly respond to requests embedded with interactional phenomena such as ellipsis and deixis.

In addition to responding to human actions through speech and vision, robots need to display proper actions to humans for smooth communication. We are investigating this issue through a museum guide robot. In this project, we place an emphasis on "personable" human-robot interaction through nonverbal behavior. In previous research, there were several museum guide robot projects, e.g., [7]. These mainly focused on robot autonomy and not on interaction with humans. Yet, we recognize that it is important to utilize nonverbal behavior in order to develop a personable and effective robot. Sidner et al. [8] have conducted a similar experiment on a guide robot designed to explain some innovative items. Bennewitz et al. [9] have recently developed a humanoid guide robot that interacts with multiple persons. This robot can direct the attention of its listeners toward objects of interest through pointing and eye gaze. In addition, Shiomi et al. [10] have done a longitudinal study on human-robot interaction at Osaka Science Museum. These studies, however, have not paid attention to the ways gestures and body movements can potentially be co-ordinated with talk in human-robot communication.

Recent research has attempted to develop effective gestures such as head movement in human-robot communication by studying human communication with a focus on gesture, head movement and eye gaze. In particular, Sidner et al. [8] developed a penguin robot and examined how users reacted toward the robot in the context of the robot explaining an exhibit; in the first condition, the robot continuously gazed toward the user, whereas in the second the robot moved its head and arms occasionally during the explanation. Under the second condition, it appeared that user attention was more greatly captured as the users appeared to respond to the robot's head movements and gaze direction by changing their own gaze and head directions. Though primarily focusing on emotion, Breazeal's study [11] suggests the importance of nonverbal interaction between humans and robots. As revealed in the above studies, research on human-robot communication has illuminated the importance of robot head movement and gesture in listener attention and response.

To investigate what proper actions robots need to display to their human recipients, we first examined the behavior of human guides through analytical methods employed in sociology. Based on these findings, we developed a museum guide robot that moves its head in particular ways. The results reveal some of the potentialities of coordinating talk and gesture in human-robot communication, arrived at through collaboration between engineering and sociology researchers.

11.2 Understanding Simplified Utterances Using Robot Vision

11.2.1 Inexplicit Utterances

As mentioned above, visual information shared by speaker and listener often allows the speaker to produce inexplicit utterances. Among such utterances are deixis and ellipsis, which may also appear in the speech interface with the helper robot [6].

Human utterances directed toward the robot are in the form of requests (i.e., utterances that get the addressee to do an action, such as "Bring the remote control"). Such utterances may consist of a verb and an object. The verb indicates the action the human wants the robot to do. The object indicates the target of the action. For each verb and object, the human may say it either definitely, ambiguously, or omit it completely. In our protocol, saying the verb or object "directly" means using a full noun or verb (e.g., remote control [N], bring [V]). Saying the verb or object "ambiguously" means using a relatively nonspecific verb (e.g., do, make), or deixis in place of a noun (e.g., this, that). Finally, "omitting" an object or verb means not saying it at all. Based on these two parts of speech (verb and object) and three ways to refer to it (definitely, ambiguously, or omitting it), utterances were classified into nine cases as shown in Table 11.1. It should be noted that the original language was

Table 11.1 Utterance classification

	Verb	Object	Example
Case 1	omitted	omitted	"Hello."
Case 2	omitted	ambiguous	"That one."
Case 3	omitted	definite	"That apple."
Case 4	ambiguous	omitted	"Make to four" (while watching television).
Case 5	ambiguous	ambiguous	"Do that."
Case 6	ambiguous	definite	"Do the red one." ("Red one.")
Case 7	definite	omitted	"Get." ("Get it.")
Case 8	definite	ambiguous	"Get that."
Case 9	definite	definite	"Get the red book."

Japanese. Here, we provide direct translations with clarifications in parentheses. Cases 2-8 were considered inexplicit utterances for the purposes of this study.

11.2.2 Information Obtained by Vision

When we interact with others, we monitor each other in order to understand what the conversation partner is doing, intending, referring to, and the like. Visual information is often crucial to understanding inexplicit utterances such as those given above. In this section we will illustrate what kinds of visual information can be used to make out the object in inexplicit utterances. We may omit an object through ellipsis or mention it ambiguously through deixis because we assume it is apparent to the listener. It is presumed that the interactants are committed to the object in some sense. Thus, the object must be something related to action. Embodied movements and other contextual features that allow for such ambiguity or omission are as follows.

1. Proximity: Objects close to the human or the robot.
2. Gaze: Objects in the line of vision of the human.
3. Pointing: Objects pointed toward by the human's arm and hand.
4. Manipulation: Objects touched or manipulated by the human or the robot.

If the object is uncertain after taking in the language and analyzing it, the system tries to detect the object based on the above features. If multiple objects have been detected, the object being pointed at is given first priority, since pointing is highly intentional. In other cases, the robot asks the human for clarification of the target object.

The robot recognizes the name of objects based on a computerized dictionary. Nouns are classified into two groups: (1) things moved frequently and easily such as 'book' and 'apple', and (2) things moved infrequently and not easily such as 'bookshelf' and 'television'. Gibson classified things perceived into five categories: places, attached objects, detached objects, persisting substances, and events [12]. Here, we employ Gibson's terminology as in our previous research [6]. According to this terminology, objects in the former group (e.g., book and apple) are considered detached objects, and those in the latter (e.g., bookshelf and television) are considered attached objects.

The system predicts the verb based on the object mentioned definitely or guessed at based on the method described above. The list of possible verbs for each object is registered in the dictionary. The default verb for detached objects is "get". If a detached object can have other possible verbs, these are supplied for each object. For example, a human may ask the robot to do various actions with a TV remote control (e.g. turn it on, raise the volume, or change the channel). For attached objects, the default action is to go to the object. When there are multiple possible verbs, we did not assign priorities in the current implementation. In such cases, the robot asks the user for clarification through speech.

11.2.3 Language Processing

We have employed ViaVoice by IBM for speech recognition. We divided speech recognition output into morphemes and parsed them using software developed at Nara Institute of Advanced Science and Technology [13]. From the parsing result, we classified sentences into the nine patterns shown in Table 11.1.

Figure 11.1 shows several examples of the parsing results. After parsing, we assigned either D (definite) or A (ambiguous) to the object and the verb. If the verb or the object is omitted (missing), M is assigned. Example 1 in Figure 11.1 is analyzed as a perfect (explicit) request, Case 9, since the result is {D, D}. Examples 2, 3, and 4 are inexplicit requests, Case 4 {A, M}, Case 8 {D, A}, and Case 3 {M, D}, respectively. After the utterance is classified, vision processes are initiated depending on the case.

11.2.4 Vision Processing

We have developed a robot system with two stereo camera pairs as shown in Figure 11.2. The lower stereo pair using IEEE 1394 cameras (DFW-V500, Sony) watches the user's face and hands, obtains the face direction (rough gaze direction), recognizes pointing gestures, and detects the objects touched or manipulated by the hands. The upper stereo pair of pan-tilt controllable cameras (EVI-D100, Sony) searches for objects along the 3-D line of the face direction or pointing direction using a zero-disparity filter (ZDF) [14]. These cameras also detect objects in front of the human and the robot.

Legend

S → Sentence
V → Verb
N → Noun
P → Noun-Pronoun
Adj → Adjective
Ot → Others (interjection, number etc.)
NP → Noun Phrase

Request Sentences

A perfect request sentence { <verb>, <object + Attribute>}
Example 1: Get me that red book {D, D}
　　　　　　S1 → V Adj N
Example 2: Make to 4 {A, M}
　　　　　　S2 → V Ot
Example 3: Get that {D, A}
　　　　　　S3 → V P
Example 4: That apple{M, D}
　　　　　　S4 → N P

Fig. 11.1 Judgment of inexplicit request patterns.

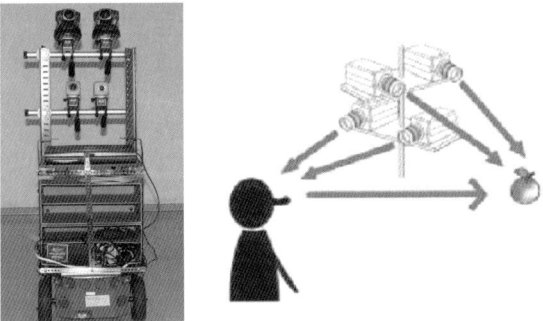

Fig. 11.2 Robot system with two stereo camera pairs.

We use the 3D human motion recognition system MARIO developed at Kyushu University [15] to detect face and hands. The system then computes the 3-D direction of the face (or arm). We consider face direction to be approximate gaze direction. The two pan-tilt controllable cameras rotate while their optical axes converge on the 3-D line indicating face (or arm) direction. The system calculates the correlation between the central regions of the two camera images. If the correlation is high, the system judges that an object exists there. Figure 11.3 shows an experimental

Fig. 11.3 Experimental
scene.

Fig. 11.4 Object detection result.

scene. Here the robot detected the electric pot that the human was looking at using
the ZDF. Figure 11.4 shows a stereo pair of images (left and center) and the ZDF
result (right). The detected object is indicated with a square.

11.2.5 Synchronization Between Speech and Vision

We also need to consider synchronization between speech and vision. Vision processes
that track the face and hands are working all the time. If a hand is raised during an
utterance, this hand motion is considered a pointing gesture. Actions of being near
and manipulating are not so fast so the robot can start the vision processes after the
analysis of utterances. Gaze, however, moves faster. Furthermore, gaze may change
(multiple times) during the course of producing an utterance. Object detection based
on the ZDF cannot work as fast as the speed of human eye movements. Even if it
could, the robot needs to determine the target object if multiple objects are detected
in multiple gaze directions. Thus, we performed an experiment to examine synchronization between speech and gaze direction.

In this experiment we put five objects in the scene. A subject sat close to the robot
as shown in Figure 11.3, asking it, "Ano (that) [object name] totte (get)." ('Get that
[object name].') The robot computed the gaze (face) direction during the period
from a little before the utterance to a little after. We utilized three participants, all
graduate students in our department. We asked them to change the [object name]
randomly and make the request. Each subject made twenty requests to the robot.

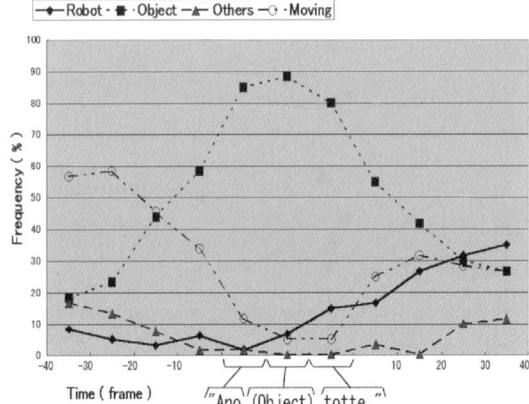

Fig. 11.5 Gaze direction changes when making a request.

Figure 11.5 shows the results. We considered gaze direction holding for more than five frames (0.17 s) to be meaningful, while others are transient. The figure shows the frequency (percentage out of 60 trials) of gaze direction during the period of each word utterance (approximately 10 frames) and during 10-frame (0.3 s) periods before and after each utterance. In the figure, gaze direction is specified by what exists there such as "Robot" and "Object" (the target object). For example, "Robot" indicates that the participant looked toward the robot. "Others" indicates objects other than the target object. "Moving" means gaze direction was changing during the period and no meaningful gaze direction was observed. The results indicate that the subjects begin to gaze toward the target object before the utterance and continue to gaze toward it when they utter its name. This is congruent with the findings of Kaur et al. on their gaze-speech input system [16].

We used explicit utterances (Case 9) in the above experiment as it proved difficult to design experiments in which subjects used inexplicit utterances in a natural way. When we asked subjects to use inexplicit utterances (as in Cases 2–8 earlier), subjects tended to fix their gaze on the target object. Experiments in which we test the use of inexplicit utterances are left for future work. Still, the above experimental results suggest how the robot can determine gaze direction when it searches for a target object. Important findings from this experiment are that primary gaze direction is toward the object or robot, and that gaze tends to begin moving toward the object before the speaker produces the utterance. Thus, we have set up the system as follows. After the utterance, the robot starts searching for an object in the gaze direction observed most frequently during the period from a little before the utterance until the end (excluding the speakers' gaze toward the robot). If multiple stable gaze directions are observed during this time period, the one around the starting time of the utterance is examined first.

11.2.6 Experiments

We performed experiments in various cases to confirm the usefulness of our approach. Here, we show an example of a dialog between a user and the robot. In this case, both participants were looking in the same direction toward two red apples. The dialog and the robot actions in this experiment were as follows.

User: "Get that."

The utterance was classified as Case 8 (verb definite, object ambiguous). The robot recognized the user's face direction, and detected two objects in the direction (Figure 11.6). The robot then verbally conveyed its current understanding status (image processing result) to the user.

Robot: "I have found two red round objects. Which should I get?"
User: "Get the left apple ."
Robot: "This one?"
(The robot shows the user the display where the target object is placed).
User: "Yes."

Figures 11.7, 11.8 and 11.9 show other typical examples. The robot was able to understand the utterances by using vision in these cases.

Although the current system was used only to conduct a small experiment with a vocabulary size of 300 words, the results show that the system can understand such inexplicit utterances as in Section 11.2.1.

Fig. 11.6 Experimental scene for an example of human-robot dialog. The objects in the user's face direction (left), and the detected objects by vision (right).

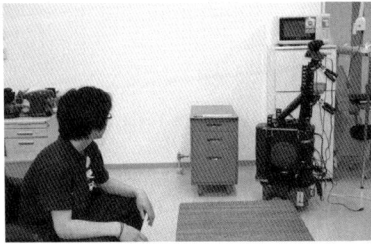

Fig. 11.7 Experimental example 1. When the user said, "Get that," the robot recognized the object in the right figure as the one referenced by 'that' because it was close to the robot and in the direction of the user's face.

 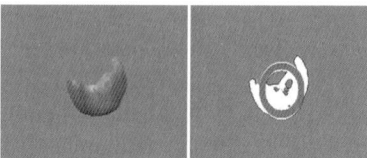

Fig. 11.8 Experimental example 2. When the user said, "Get that," while pointing toward the object, the robot recognized the object in the rightmost figure as the one indicated by 'that' because it was being pointed at.

Fig. 11.9 Experimental example 3. When the user said, "Throw this," the robot recognized the object in the right figure as the one indicated by 'this' because was being held by the user.

The robot may find multiple objects associated with the human's various actions. In the current implementation, we have not specified a hierarchy among them except in the case of pointing. The robot asks the human through speech in such cases. More detailed language analysis and/or the use of other visual information may solve this issue. This is left for future work.

In relation to the above experiments, we used uncluttered simple scenes in order to confirm that the mechanism can interpret and respond to inexplicit utterances through visual information. Currently, the robot may have trouble detecting objects in cluttered scenes. In order to deal with such situations, we are working on an interactive object recognition [17, 18, 19], in which the robot asks the user to provide further information about the object when it cannot immediately detect it.

11.3 Communicative Head Gestures for Museum Guide Robots

In this section, we discuss our research exploring proper actions for smooth human-robot communication. We are investigating this issue through developing a museum guide robot that moves its head appropriately while providing an explanation of exhibits.

11.3.1 Observations from Guide-Visitor Interaction

Before attempting to develop a guide robot, we observed how human guides behave in two situations. We performed both experiments in our lab. In this experiment, one guide explained an exhibit of the history of roof tiles in ancient Korea. The guide did four explanations for fifteen minutes each to four visitors (one at a time), and two explanations for thirty minutes to two pairs of visitors. The guide was a researcher on the exhibit and the visitors were university students. We video recorded the experiments. Figure 11.10 shows the experimental scene.

We performed the second experiment using an exhibition of photographs introducing Thailand. The guide was the photographer himself, who explained to three different visitors and a pair of visitors, each for about thirty minutes. The visitors were university students. We recorded the experiments with video cameras.

Upon reviewing the video segments, we extracted 136 instances in which the guides clearly turned their heads toward the visitor. Table 11.2 summarizes the instances of head movements. The guides made frequent head movements at transition relevance places (TRPs) -places in the talk where it is most appropriate for the listener to take a turn [20] such as at the completion of a sentential unit.

Figure 11.11 shows an example. The guide is on the left and the visitor is on the right. The guide was explaining the process of making roof tiles. The following excerpt shows talk and gaze direction by both the guide and the visitor. In the left figure, the guide faces the visitor, which comes toward the end of a sentential unit.

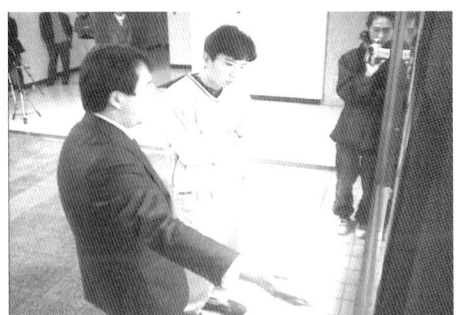

Fig. 11.10 Guide scene at Saitama University.

Table 11.2 Number of cases guides turned their heads in the two experiments

	Number of occurrences
TRP (transition relevance place)	61
When saying key words with emphasis	14
When saying unfamiliar words or citing figures	6
When using deictic words such as "this"	26
With hand gestures	41
When the visitors asked questions	12

Total 136 times: counted multiple if multiple conditions are satisfied.

Fig. 11.11 Examples of the TRP case. Left: The guide (left) turns toward the visitor (right). Right: The guide turns back toward the exhibit to proceed to the next explanation.

Fig. 11.12 The guide points while saying, "Here."

Excerpt 1 (Gaze direction and gesture: X= gaze toward the exhibition, P=pointing, V=gaze toward the visitor, N=nodding)

```
01 G:   De ma: kore ga kansei ban to, kouiu katachi de ma: kawara ga dekirun desuNE:.
        so well  this   final form and  this form    so well   tile    made is
        'So, this is the final form, and the tile is made into this form.'
G:      XXXXPPPPPPPPPPPPPPPXXPPPPPPPPPPPPPPPPPPPP....VVVVVVVVNNV.....
V:      XXXXXXXXXXXXXXXXXXXXXXXXXXXXXXXXXXXXXXXXXXXXNNNNNNN
```

In this excerpt, the guide is pointing toward the exhibit while explaining the tile. In particular, when the guide says the deictic terms "this" (kore) and "like this" (kouiu), he is pointing toward the exhibit. The beginning element of this turn "so" (de) marks this sentence as the final element of explanation. When he says "the tile is made", the guide starts to move his head slightly toward the visitor. This movement allows him to check whether or not the visitor is displaying understanding at this point. This sentence ends with a slight rise in intonation, in particular at "made" (dekirun), indicating the turn is coming to completion. At this point, both the guide and the visitor start nodding at the same time. These movements display a certain degree of mutual understanding.

In addition to TRPs, guides also turned their heads toward the visitor when saying key words. Figure 11.12 shows one example from the second experiment. In this example, the guide was showing some pictures of small shrines in Thailand. In excerpt 2, the guide turns his head toward the visitor as he says the name of the ghost (Pi), a keyword.

Excerpt 2

01 G: Koko- you wa, kokoni (0.3) e:::to (0.8) you wa .hhh tai no yuurei.
       here   in short here   well   in short   Thailand ghost
       'Here   in short here   well   in short   ghost in Thailand'
G: XPPPPPPPPPPPPPPPPPPPPPPP PPPPXXXXXXXXXXXXXXXXXXX
V: XXXXXXXXXXXXXXGGGGXXXXXXXXXXXXXXXXXXNNNNX

02 G: Pi: toiu   [noga oru]   pi::.
       "Pi" called   is   exist   "Pi"
       ' "Pi" is here "Pi" '.
G: ...VVVVVVVVVVVV...

03 V: [pi::?]
V: ...GGGGGGGGGGGGNNNNXX

04 G: .hhh de, sorega sunderu.
       and   that   live
       'And they live (in there).'
G: ...XXXXXXXXXXXXXXXXXXXX
V: XXXXXXXXXXXXXXXXXXXXXX

At line 1, the guide is pointing and gazing toward the picture while explaining that the ghost resides inside the small shrine. At line 2, he marks the term "Pi" (name of the ghost) as new or unfamiliar information by prefacing it with "toiu (it is called)". While saying "Pi", the guide turns his head toward the visitor. The guide's gaze indicates an attempt to check the visitor's understanding. At this point, the visitor also starts looking at the guide, and repeats the term "Pi" (with rising intonation) at line 3. This repetition functions as a check for understanding. The visitor starts nodding as the guide confirms the term by repeating it. This exchange and mutual gaze during the exchange clearly display that the visitor registers the term as something new and significant in the guide's explanation.

The guides often turned their heads and also made hand gestures when using deictic words. These two actions typically appeared simultaneously. Again, in excerpt 2, and shown in Figure 11.12, the guide points at a certain part in the picture at line 1, while producing the deictic word "here" (koko).

These experiments show that head movements and other embodied actions occur at fairly predictable places within the talk of exhibit guides. In employing robots to do the work of guides at a museum, it may be important for a robot to deploy nonverbal behavior at interactively appropriate points to create a more naturalistic interaction in general and a more personable robot in particular.

11.3.2 Prototype Museum Guide Robot

Based on the above findings from our guide experiment, we developed a prototype museum guide robot that moves its head while explaining exhibits. Figure 11.13 shows a photograph of the robot. The robot has two pan-tilt-zoom cameras

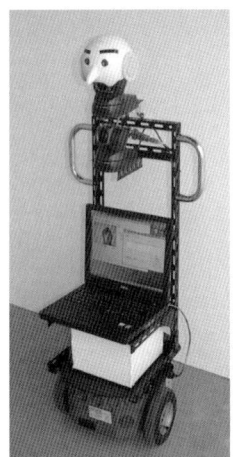

Fig. 11.13 Guide robot. The robot consists of a mobile robot, Pioneer II by Activ-Media, a laptop PC, and two pan-tilt-zoom cameras.

(EVI-D100, Sony). We attached a plastic head on the upper camera and used the pan-tilt mechanism of the camera to move the head. We did not use the images of the upper camera in the current implementation. The robot uses the images of the lower camera to make eye contact and to observe the visitor's face.

Visitor eye contact toward the robot may function as request for help. When a visitor stands close to an exhibit and makes eye contact, the robot approaches the person and asks, "May I explain this exhibit?" If the visitor answers "Yes", the robot begins explaining the exhibit. The actual eye contact process is as follows. The robot pans around with its lower camera to find a visitor who is gazing toward the robot. If it finds such a visitor, it turns its body toward him/her. If he/she is still gazing toward the robot, the robot assumes that the visitor might like an explanation of the exhibit. This eye contact process is the same as the one in our eye contact robot [21, 22], except that the current robot has a head-shaped figure instead of a Computer Graphic head.

Now let us briefly describe the face image processing method used for eye contact. Our robot first searches for face candidates with the zoomed-out camera. When a candidate is detected, the camera zooms in. The robot then examines detailed facial features.

The candidate face regions can be detected in the images with a wide field of view. First, skin color regions are extracted. Then, small regions and greater elongated regions are removed. Inside the remaining regions, subtraction between consecutive frames is computed. The largest region among those where the sum of absolute values of the subtraction exceeds a given threshold is considered a face candidate. Figure 11.14 illustrates an example of a face candidate. The pan, tilt, and zoom of the camera adjust so that the candidate region can be taken large enough to examine facial features. Experiments show that it can detect human faces indoors at a distance of six meters.

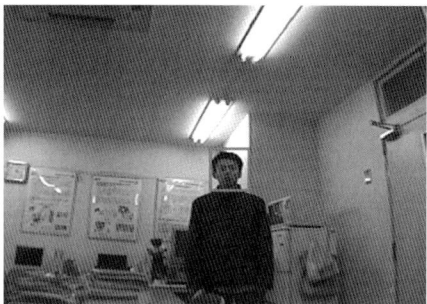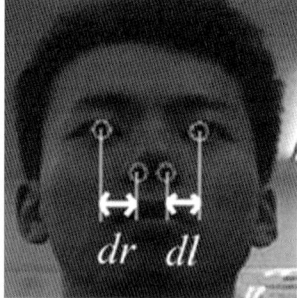

Fig. 11.14 Face image processing. Left: Face candidate. Right: Face direction computation.

The system detects the eyes (pupils) and the nostrils in the zoomed-in image. We use the feature extraction module in the face recognition software library by Toshiba [23]. Then, the system measures the horizontal distance between the left pupil and the left nostril dl and that for the right side dr as shown in Figure 11.14. From these two values it determines the direction of the gaze (face). The robot does not actually need to compute the accurate direction. It only needs to determine whether or not the person is looking at the robot. Since the camera has turned in the human's direction, the frontal face must be observed if the human is looking at the robot's face. If the ratio between dl and dr is close to 1, the human face can be considered to be facing toward the robot. This same computation process is used while the robot is explaining the exhibit.

The robot explains the exhibit using synthesized speech. While speaking, the robot turns its head toward the visitor at similar points identified in the human guide experiment described above. There are two types of head motions: predetermined and online. The observations from the guide-robot interactions showed that human guides often turn their heads at certain points during the explanation. We manually inputted annotation marks for the robot to turn its head at such positions in the text of the explanation. We call such cases predetermined ones. In the current implementation, we chose the following points based on our earlier guide experiments.

-TRP 1: At the end of a certain explanation.
-TRP 2: When the robot asks a question.
-When the robot says a key word or unfamiliar word.
-When the robot uses a deictic word to refer to something.

In the online cases, the robot reacts to the visitor. In particular, the robot turns its head when it sees the visitor turning his/her head toward the robot. The robot is able to do this since it continuously monitors the visitor's face direction with the lower camera. In response to the visitor's head movement toward the robot, the robot turns its head toward the visitor and says, "Do you have any questions?"

The robot can obtain the movement of the visitor's face when it turns its head in predetermined cases. This information can indicate the visitor's response to the explanation by the robot. The robot should be able to modify the explanation depending on the visitor's response. However, the current robot is not yet able to do this. This is a task for future work. In addition, the current robot cannot answer questions if the visitor asks them. As a result, we have not implemented the online head turning in the experiments described below.

11.3.3 Experiments at a Museum

We organized an interactive art exhibition using magnetic fluid by Sachiko Kodama and Minako Takeno at Science Museum, Tokyo, from December 3 through 17, 2005. We demonstrated our robot on December 12 and performed experiments.

Sixteen visitors agreed to participate in our experiments (14 females, 2 males, ages 20–28, students and office clerks). When a visitor stands near the artwork Morphotower, and makes eye contact with the robot, the robot comes close to the visitor and explains the work. The robot explains the work in two modes: the proposed mode in which the robot turns its head to the visitor at predetermined points and the fixed mode in which the robot continuously gazes toward the exhibit without turning its head. In the former mode, however, the robot does not use online head turning, because the robot cannot answer the visitor when the visitor asks a question in the current implementation.

Eight participants engaged in the fixed mode followed by the proposed mode (Group A). The other eight participants did so in the reverse order (Group B). We allowed about a half an hour interval between the two modes. The participants were asked to look around the museum during the interval and not to observe the experiments by the other participants. We did not tell the participants the differences between the two modes. We videotaped the experiments. Figure 11.15 shows an experimental scene.

After the experiments, we asked the participants which presentation mode they would prefer if the robot were to provide an explanation again. For the participants of Group A, six preferred the proposed mode and two the fixed mode. These numbers, 6 and 2, are the same for the participants of Group B. The results suggest that while viewing the museum exhibits with a robot guide, visitors prefer robot head movements to no head movements, although the evidence is not decisive since the

Fig. 11.15 Robot experiments at Science Museum, Tokyo.

number of participants was small, and gender and ages of the participants did not vary much.

As a quantitative evaluation, we examined when and how often participants turned their heads toward the robot. In the proposed mode, the robot turned its head 7 times for each trial at predetermined points as follows.

1. When the robot approaches the visitor, the robot gazes toward the visitor, and then turns its head toward the work while saying that it will now explain the exhibit. At this time, head-turning direction is different from the other six points where the robot turns its head from the exhibit to the visitor.
2. The robot emphasizes the key word 'magnetic fluid'.
3. The robot uses the deictic word 'this'.
4,5,7 TRPs: The robot finishes an explanation.
6. TRP: The robot asks a question.

Figures 11.16 and 11.17 show the percentages of the participants moving their heads around each predetermined point for Group A and for Group B, respectively. In these figures, the horizontal axes indicate the time scale with the seven

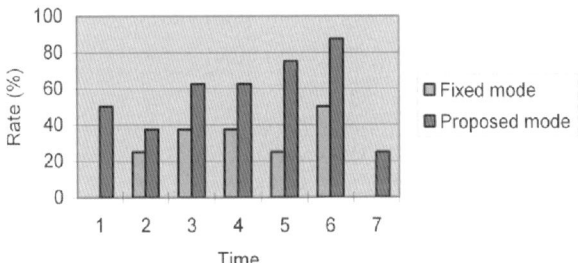

Fig. 11.16 Rate of participants moving their heads when robot turned its head. (For participants who tried the fixed mode first (Group A).)

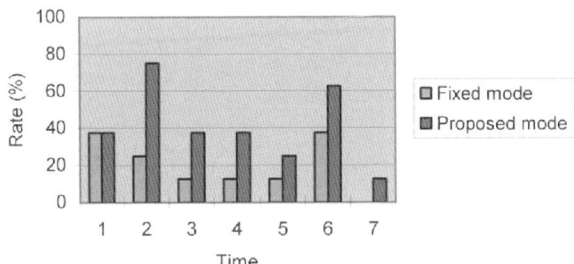

Fig. 11.17 Rate of participants moving their heads when the robot turned its head. (For participants who tried the proposed mode first (Group B).)

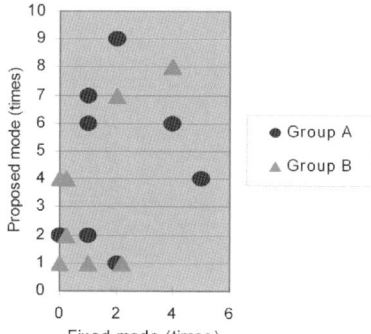

Fig. 11.18 Number of head movements of each participant in both fixed and proposed modes.

predetermined points. At the first predetermined point, both figures show the percentages of participants who turned their head from the robot to the exhibit in response to the head movement of the robot. At other predetermined points, the figures show the percentages of participants who turned their heads from the exhibit to the robot.

Both figures show that the percentages of participants' heads movements increase significantly in the proposed mode ($p < 0.01$, paired t-test). This significance appears in both groups A and B, which suggests that the increase in head movement is not dependent upon the order of the two trials. In the fixed mode, the participants move their heads toward the robot at the point where the content of the explanation solicits the participants' attention toward the robot such as when the robot asks a question.

Figure 11.17 shows that the participants who experienced the proposed mode first gradually decrease their number of head movements, even though they are moving their heads in the beginning of the fixed mode. The participants also turn their heads toward the robot at the sixth point where the robot asks a question.

In Figure 11.18, the horizontal axis indicates the number of head movements in the fixed mode, and the vertical axis shows the same in the proposed mode for each participant. This figure also shows that the participants' head movements increased in the proposed mode.

We recognize there is a possibility that it may be a natural response for humans to turn their heads toward the robot when the robot turns its head toward them, and the larger number of head movements of the participants may not necessarily mean that the robot in the proposed mode is more user-friendly or personable. However, considering the finding that 12 participants out of 16 conveyed a preference for the proposed mode, we suggest that it may be effective for guide robots to turn their heads toward the visitor(s) at certain points while explaining exhibits.

11.4 Conclusion

We have presented two robot systems through which we have shown the importance of vision and action for human-robot communication.

The first robot can understand simplified utterances through computer vision. In daily conversation, we often omit some information or mention ambiguously things we assume the listener can identify through vision. We have presented a robot that can appropriately respond to user requests based on speech and vision. The robot tracks human gaze direction, detecting objects in its direction. It also recognizes other human actions such as pointing. Based on visual information, the robot understands simplified utterances that contain ellipsis and deixis.

The second robot suggests that proper head gestures can increase the engagement of humans with the robot, and possibly lead to a more enjoyable experience for visitors. Face and head movements play an important role in human communication. We have presented a museum guide robot that moves its head in an attempt to communicate with humans. By analyzing the behavior of human guides when they explain exhibits to visitors, we used this information to develop a robot system that turns its head at predetermined places and in response to human behavior.

These robots are still in their early stages of development. To understand human utterances (and intentions) more fully, there are other actions the robot has to recognize. Even though our current robot can recognize some mundane actions, the robot cannot yet detect target objects in complex environments. We need to improve the capability of object recognition. In cases where the robot fails to recognize objects, we are working on interactive object recognition [17, 18, 19]. The robot asks the user to give information about the object when it cannot detect it. The research presented in this paper can be called "Vision for Communication," whereas this interactive object recognition can be called "Communication for Vision."

In terms of the guide robot, we have used the mechanical part of a pan-tilt camera to move the robot's head. Thus, the robot cannot move its head as fast and as subtly as humans. We have recently obtained Robovie-R ver.2 [24]. We are now working on implementing the head-turning method for the robot. We will perform more experiments with the robot to confirm the usefulness of head gestures. We will also examine the effects of other actions such as hand gestures and body movements. We also plan to investigate ways to modify the robot's explanation in accordance with the observation results of the museum visitors.

Acknowledgements This work was supported in part by the Ministry of Internal Affairs and Communications under the Strategic Information and Communications R&D Promotion Program, and by the Ministry of Education, Culture, Sports, Science and Technology under the Grants-in-Aid for Scientific Research (KAKENHI 14350127, 18049010).

References

1. Graf, B. and Hägele, M.: Dependable Interaction with an Intelligent Home Care Robot. Proc. ICRA 2001 (2001) 21–26
2. Seabra Lopes, L. and Teixeira, A.: Human-Robot Interaction through Spoken Language Dialog. Proc. IROS 2000 (2000) 528–534
3. Schiehlen, M.: Ellipsis Resolution with Underspecified Scope. Proc. ACL 2000 (2000) 72–79

4. Watanabe, M., Masui, F., Kawai, A. and Shino, T.: Conversational Ellipsis and Its Complement. Trans. IEICE 2000 SP2000-99 (2000) 31–36 (in Japanese)
5. Grice, H.P.: Logic and Conversation. Harvard University Press (1975) 120–150
6. Yoshizaki, M., Nakamura, A., and Kuno, Y.: Vision-Speech System Adapting to the User and
   Environment for Service Robots. Proc. IROS 2003 (2003) 1290–1295
7. Nourbakhsh, I., Kunz, C., and Willeke, T.: The Mobot Museum Robot Installations: A Five
   Year Experiment. Proc. IROS2003 (2003) 3636–3641
8. Sidner, C.L., Lee, C., Kidd, C.D., Lesh, N., and Rich, C.: Explorations in Engagement for
   Humans and Robots. Artificial Intelligence, Vol. 166 (2005), 140–164
9. Bennewitz, M., Faber, F., Joho, D., Schreiber, M., and Behnke, S.: Towards a Humanoid
   Museum Guide Robot That Interacts with Multiple Persons. Proc. 2005 5th IEEE-RAS Int.
   Conf. on Humanoid Robots (2005) 418–423
10. Shiomi, M., Kanda, T., Ishiguro, H., and Hagita, N.: Interactive Humanoid Robots for a Sci-
    ence Museum. Proc. HRI2006 (2006) 305–312
11. Breazeal, C.: Emotion and Sociable Humanoid Robots. International Journal of Human-
    Computer Studies, Vol. 59 (2003) 119–155
12. Gibson, J.J.: The Ecological Approach to Visual Perception. Houghton Mifflin (1979)
13. Matsumoto, Y., Kitauchi, A., Yamashita, T., Hirano, Y., Matsuda, H., Takaoka, K., and Asa-
    hara, M.: Japanese Morphological Analysis System ChaSen Version 2.2.4 Manual. Nara In-
    stitute of Science and Technology (2001) (in Japanese)
14. Coombs, D., and Brown, C.: Real-time Binocular Smooth Pursuit. IJCV, Vol. 11, No.2 (1993)
    147–164
15. MAlib development team. http://www.malib.net/. Cited 8 Sept 2007
16. Kaur, M., Tremaine, M., Huang, N., Wilder, J., Gacovski, Z., Flippo, F., and Mantravadi,
    C.S.: Where is "it"? Event Synchronization in Gaze-Speech Input Systems. Proc. ICMI 2003
    (2003) 151–158
17. Kurnia, R., Hossain, M.A., Nakamura A., and Kuno, Y.: Generation of Efficient and User-
    friendly Queries for Helper Robots to Detect Target Objects. Advanced Robotics, Vol.20,
    No.5 (2006) 499–517
18. Hossain, M.A., Kurnia, R., Nakamura, A., and Kuno, Y.: Interactive Object Recognition Sys-
    tem for a Helper Robot Using Photometric Invariance. IEICE Trans. Inf. Syst., Vol.E88-D,
    No.11 (2005) 2500–2508
19. Hossain M.A., Kurnia R., Nakamura A., and Kuno. Y: Interactive Object Recognition through
    Hypothesis Generation and Confirmation. IEICE Trans. Inf. &. Syst., Vol.E89-D, No.7 (2006)
    2197–2206
20. Sacks, H., Schegloff, E. A., and Jefferson, G.: A Simplest Systematics for the Organization
    of Turn-Taking for Conversation. Language, Vol. 50 (1974) 696–735
21. Miyauchi, D., Nakamura, A., and Kuno, Y.: Bidirectional Eye Contact for Human-Robot
    Communication. IEICE Trans. Inf. Syst., Vol. E88-D, No.11 (2005) 2509–2516
22. Miyauchi, D., Sakurai, A., Nakamura, A., and Kuno, Y.: Active Eye Contact for Human-
    Robot Communication. Extended Abstracts CHI 2004 (2004) 1099–1102
23. Fukui, K., and Yamaguchi, O.: Facial Feature Point Extraction Method Based on Combination
    of Shape Extraction and Pattern Matching. Systems and Computers in Japan, Vol.29, No.6
    (1998) 49–58
24. ATR Intelligent Robotics and Communication Laboratories,
    http://www.irc.atr.jp/productRobovie/robovie-r2-e.html. Cited 8 Sept 2007

Index

Printed in the United States of America